SAFETY
IN WORKING
WITH CHEMICALS

DATE DUE			
Dec 9 79			

SAFETY IN WORKING WITH CHEMICALS

MICHAEL E. GREEN   AMOS TURK
Department of Chemistry
The City College of the
City University of New York

MACMILLAN PUBLISHING CO., INC.
New York
COLLIER MACMILLAN PUBLISHERS
London

542
G82s
110911
Sept. 1979

Macmillan Publishing Co., Inc.
866 Third Avenue, New York, New York 10022

Collier Macmillan Canada, Ltd.

Library of Congress Cataloging in Publication Data

Green, Michael E
 Safety in working with chemicals.

 Includes index.
 1. Chemical laboratories—Safety measures.
I. Turk, Amos, joint author. II. Title.
QD51.G73   542   78–5122
ISBN 0–02–346420–8

Printing: 1 2 3 4 5 6 7 8   Year: 9 0 1 2 3 4 5

Preface

In spite of the gathering evidence everywhere that chemical laboratory workers, including students, are subjected to greater environmental risks than the general population, safety instruction continues
to be a haphazard activity. Often it is limited to the first "indoctrination" hour of the first laboratory period, when the student is only
dimly aware of what will be going on during the remainder of the
course. Furthermore, the usual emphasis is on immediate dangers,
such as that of fire, while the hazards of chronic exposures are
relegated, at least implicitly, to a secondary role.

It is tempting to recommend that separate courses in laboratory
safety be instituted, but we are aware that such a proposal is unrealistic for an already crowded curriculum. Besides, we do not believe
that "safety" should be separated from "chemistry," either in instruction or in concept. Accordingly, it is our recommendation
(elaborated in appropriate chapters of the book) that existing laboratory courses accommodate safety instruction as an integral part of
all the prescribed experiments.

We have tried to do more than write a set of rules or a safety
"outline." However, this book is not an encyclopedia. Instead, it
aims to *teach* the subject of chemical safety in a way that will
provide understanding of the fundamental concept of safe practice.
Its ultimate object is to contribute to the reduction of injuries and
illness.

The book can be useful in a variety of ways. It can serve as a text for a short course in laboratory safety. It can be used as an indoctrination to safe practice for newcomers to any chemical facility —teachers, technicians, researchers, and students. Certainly it should be in the hands of all members of chemical safety committees.

The authors will appreciate all comments and suggestions regarding the subject matter of the book. All communications will be answered.

Michael E. Green
Amos Turk

Contents

1

Introduction

1.1 CHEMICAL HAZARDS—ACUTE AND CHRONIC

Once, when one of us (M.G.) was travelling in another country, he tried to communicate with a student across a language barrier. There was only one way he was able to explain to the student that he was a chemist. After pointing to himself, he slowly mimed pouring one liquid into another, following this by shouting BOOOMM!! Instant comprehension resulted.

Chemists have become all too notorious for their association with unwanted and accidental explosions. It is also well known that many of the substances chemists work with are toxic, some very much so. Until recently, the chronic effects of most of these substances were not well recognized, although a few chronic toxins have been known for many years. As early as 1775, cancer of the scrotum of chimney sweeps was described by Dr. Percival Pott. Chimney sweeps removed their clothing so as not to spoil it when working. The polycyclic aromatic hydrocarbons present in soot were presumably responsible for the cancer, although Pott, writing before Dalton, had no way of knowing this. Mercury, too, has long been known to be a chronic poison.

It is now known that most cancers are environmental in origin, with estimates by recognized authorities running as high as 90 per cent. This of course includes occupationally caused cancers, which are believed to be quite numerous in this country. It is officially estimated that about 15,000 workers are killed by occupational accidents each year. However, national surveys indicate that if chronic diseases, including cancer, were included the annual occupational death toll would be closer to 100,000, and serious illnesses approach four times that number.

It was, in part, data like these that led to the passage of the Occupational Safety and Health Act in 1970. This Act, effective April 28, 1971, established the Occupational Safety and Health Administration (OSHA) and the National Institute of Occupational Safety and Health (NIOSH). OSHA is charged with enforcing the law, which states that every worker in the United States has the right to a safe workplace. (The actual effectiveness of the law in its first 6 years is open to question, principally because of inadequate enforcement, but that is another story.) OSHA has among its tasks the setting of standards for exposure to chemicals in the workplace, as well as the enforcement of these standards.

Workplaces include universities and laboratories, except for "governmental" institutions (meaning not only federal agencies and national laboratories, but state universities and state agencies, public high schools, local government facilities, and so on). However, in 1975, President Ford signed Executive Order 11807, extending inspections under the Act to federal employees, both civil and military. Some municipal and state workers are beginning to come under the effective protection of the same standards.

NIOSH is one of the National Institutes of Health (though based in Cincinnati, not Bethesda). It was established mainly to provide the research upon which standards could be based. Chemists should know that NIOSH has issued a list of 1500 suspected carcinogens, many of which are rather common substances (see Chapter 5). Chemists are therefore likely to be at risk from more different substances than any other group of workers (except, perhaps, for those who fill the bottles that will be labelled *"Dangerous—Potential Carcinogen,"* or equivalent). OSHA has only begun to inspect laboratories, and although few have yet been fined, its enforcement procedures could become more vigorous soon.

However, with or without inspections, it is worth noting that chemists apparently do get cancer at a rate approximately 25 per cent higher than that of the general population. According to a report to a Senate committee in 1971, it was predicted that about

50 million of the 200 million Americans then living would get cancer, and 34 million would die of it. Combining these figures suggests that exposure to carcinogens would cause about 21 per cent of chemists to die of cancer, compared with 17 per cent of the general population.

Even established standards are not necessarily sufficient to protect against cancer or other chronic injury; this is especially true of the older standards. New standards are being issued as new information becomes available, leading always in the direction of a tighter standard. For example, the benzene standard has dropped from 10 parts per million (ppm), averaged over 8 hr, and 50 ppm for 15 min, to one tenth these amounts. Vinyl chloride was found to be a carcinogen, and its allowable 8-hr average concentration dropped from 500 ppm to 1 ppm. Compliance with the new standards often requires considerable improvement in safe practice. Sometimes it is not easy to comply with even older standards. The permitted *ceiling* concentration for hydrogen sulfide, for example, is 20 ppm, with a 10-min excursion (once only, if no other exposure occurs) to 50 ppm.

OSHA is unable to cover most chemistry laboratories, either because of lack of inspectors or other budgetary problems, or because many laboratories are "governmental." Therefore many chemists will spend much of their careers in situations in which they are called upon to be their own safety experts, even though not directly subject to legal checks. However, teachers may be subjected to lawsuits.*

1.2 RESPONSIBILITY

Exposure to toxic chemicals is not the only laboratory hazard. The unexpected thunderclap of an explosion can be anything from distressing to disastrous; if it happens in a laboratory under a teacher's supervision, it can also mean a lawsuit. It may in addition mean severe injury, such as loss of an eye.

Faculty are responsible for more than their students' immediate safety. Attitudes toward laboratory health and safety, which are likely to be carried on long past school years, are formed in student

---

*See "The Personal Liability of Chemical Educators," *J. Chem. Educ.*, **54**:134 (1977).

laboratory courses. Obviously, these attitudes are important for future chemists. They are probably at least as important for another large group of students in undergraduate chemistry laboratories, the premedical students. Physicians, faced with an undiagnosed disease, all too often fail to look into the possibility of an occupational cause, at least until quite late in the treatment. Given the prevalence of occupational disease, it would be well if future physicians were made aware of the possibilities when they are themselves first exposed to the risk. (Even medical schools have tended to give relatively little attention to this field.)

This book should be valuable for students in undergraduate chemistry laboratory courses. In addition, the book provides a convenient manual for chemists who work at the laboratory bench and must be aware of the range of hazards they face, as well as for those who find themselves with responsibility for students, technicians, other chemists, or other workers. This is not an exhaustive treatise, obviously. Some general industrial hazards, such as lifting heavy weights, are omitted entirely. Principles and methods are described, and enough practical information is included to make the book directly useful in many circumstances. There are many other works dealing with laboratory safety, occupational health and safety in general, hazardous chemicals, and so on, as well as sources for obtaining new information in this rapidly changing field. Some of these are listed in the bibliography section at the end of the book. Each has value, and may at one point or another prove useful.

Obviously, not everything can be discussed in a book of this length. Many substances are referred to as "carcinogenic" without a serious attempt to separate strong from weak carcinogens. Given the present state of our knowledge, this would have been a fairly hopeless task. Furthermore, cost estimates are not given for such modifications as ventilating a laboratory or for doing the clinical chemistry necessary to maintain proper health checks on those exposed to a variety of chemicals. There were various points at which we could only say that certain steps would have to be taken, and recommend that expert advice be sought for the particular situation. There are some problems for which solutions cannot be stated in a few universally applicable rules; we hope that we have at least been successful in pointing out the existence of these problems.

One serious impediment to an adequate safety program is cost.

Money used for safety is unavailable for teaching or research. The problem is made more serious by the competitiveness of research. It seems likely that if chemists *as a group* insist that safe procedure be considered normal procedure, then safety equipment, supplies, and training could be funded by federal grants and supported by university administrations. Individual initiative on safety is less likely to be effective in a very competitive atmosphere. (This should not be construed as an excuse for inaction on safety but rather a statement of the necessity for cooperative action.)

We believe that most chemists will agree that increased attention to laboratory health and safety is now required; it is our hope that this book will aid in the solution of some of the problems raised.

2

Basic Laboratory Precautions

There are certain basic rules of safe procedure that all laboratories, particularly student laboratories, should follow. Most of these procedures, or rules, are known to all chemists. Unlike some of the information in the later chapters, these rules have generally been accepted for some time as being fundamental to the safe operation of a laboratory. Most are in fact discussed early in freshman chemistry. We begin with these precautionary rules.

2.1 HANDLING GLASSWARE

The most common accidents in student laboratories are cuts and burns resulting from handling glassware. It seems inevitable that in the first laboratory period of a freshman class, someone will pick up a piece of hot glass. Fortunately, the results are usually more painful than serious, and the error is not usually repeated by the same individual. To minimize the number of students who make this mistake, it is helpful to *instruct* students to place a wire gauze near their burner before starting and put the freshly bent or fire-polished glass on it for several minutes after it is heated. If the air an inch above the glass feels warm to the palm of the hand, the glass is still too hot to touch.

More serious are the cuts resulting from broken glass. Students should be told to fire-polish freshly cut ends of pieces of glass. They should also be told how to insert glass tubing into rubber stoppers. The steps are

1. Lubricate the hole in the stopper with water or glycerine.
2. Hold the glass near the end to be inserted, thus minimizing the torque.
3. Hold the glass with a towel or rag to protect the hand in case the glass breaks.

To remove glass from a stopper, wet the handle of a file with glycerine, then work the lubricated file handle between the glass and rubber. As the file is twisted, it grips the rubber, but slips on the glass, thus working in the lubricant. After this, the glass can be removed.

2.2. AVOIDING POISONS

One of the main routes of entry for toxins is the mouth. This route is not common for laboratory workers; however, it is necessary to insure that no food is in the vicinity of an area where toxic substances, especially chronic poisons, may be found. It has been determined that even when a bottle from which a liquid has been poured is resealed, enough of the liquid remains on the threads of the cap, beyond the seal, to be a problem. If the food is fatty or oily, it will dissolve fatty or oily vapors. For similar reasons, food or beverages must not be left in laboratory refrigerators.

Likewise, smoking in laboratories should be ruled out because of the risk that the cigarette might become contaminated and carry toxic substances to the mouth, even if it were not ruled out for other reasons. Another related practice that has fortunately fallen into disuse is testing chemicals by taste or odor. Finally, no one with bare feet may be allowed in the laboratory. All of these practices are so grossly unsafe as to be flatly forbidden.

2.3 HOUSEKEEPING

Keeping the laboratory neat is a matter of survival. Loose electrical wires, spilled chemicals, water hoses running across aisles (other than properly installed plumbing), and similar avoidable hazards, are invitations to disaster.

a / Student Laboratories

In student laboratories, assuming the chemicals used are not of the order of toxicity that requires extraordinary precautions (for example, carcinogens—see Chapter 5), students should be instructed to clean up any spills they cause, whether at their own benches or in common areas such as near balances or in hoods. This practice is particularly necessary for volatile substances, especially if they are toxic or flammable. Volatile toxic substances should be used only in the hood; however, spills may occur outside the hood in any case. Disposable gloves should be available for use in cleaning up these substances. At the end of the lab period, it will generally be necessary to wipe off bench tops, make sure reagent bottles have been capped and returned to their appropriate location, and so on. Students should have specific assignments, each covering a section of the laboratory; it is the responsibility of the assigned student to leave his or her section in a clean, nonhazardous condition.

Other aspects of housekeeping, such as clear aisles and safe connections, are the responsibility, at least in part, of the person in charge of the laboratory. If a laboratory is being designed, a layout that places services close to where they will be used will reduce the amount of tubing, wiring, and other connections that must be run to distant locations.

b / Research Laboratories

Considerations similar to those described in part (a) apply to research laboratories, although more toxic chemicals may be in use. The researcher in these laboratories is responsible for maintaining a neat working area. Aisles must still be clear; access to safety showers, emergency exits, and other emergency apparatus must be unobstructed.

The general rules for housekeeping are well known, but it is necessary to apply them. Actual procedures for monitoring housekeeping will be discussed in detail in Chapter 7. Controls to insure that housekeeping requirements are met will be dealt with in Chapter 8. For handling carcinogens and other extremely toxic materials, procedures similar to those recommended when radioisotopes are in use should be followed (see Chapters 5 and 6).

2.4 HORSEPLAY

One of the worst laboratory incidents one of us has seen occurred in a high school laboratory. One student told another to check the odor of a test tube; the test tube was generating chlorine gas. The second student did recover. Eventually, so did the teacher. Although there is little excuse for using chlorine outside of a hood, or using it at all in a high school lab, it is crucial that no games be played in any laboratory. The dangers of spilled acid, fire, or similar consequences of thoughtless behavior are too great to be acceptable. Teachers must warn students of the possible consequences of tricks, at least in high school and freshman laboratories. Fortunately, the problem rarely arises in research laboratories.

2.5 LONG HAIR AND TIES

Hair of greater than shoulder length may be a hazard, especially if open flames are in use or if many students are working close together. The hair may be tied above the head. This precaution also applies when one works with machine tools. Ties can equally be a problem and should either be removed or worn firmly clipped, preferably behind a lab coat.

2.6 WORKING ALONE

If someone is working in a research laboratory outside of regular hours, especially alone, it is essential that someone else be aware that the laboratory is occupied. If there is any particular hazard (acute toxicity, fire, or explosion), no person should work completely alone. Furthermore, if a particular hazard is anticipated, the necessary material or equipment to deal with it (antidote, respirator, and so on) should be prepared in advance and be available. In addition someone should be nearby who knows how to use or administer the material or use the equipment. Since most prospective hazards are not easily anticipated, a telephone, on which are listed necessary emergency numbers (fire, medical), should be in the immediate vicinity.

Even when the work is believed to be relatively nonhazardous (for example, low-voltage electronics), it is wise for someone to be on the same floor and check in regularly.

2.7 PIPETTING

The most important rule is: "never by mouth." For reasons discussed in Section 2.2, nothing in the laboratory should be put in the mouth. Although it should be obvious that one does not pipette extremely toxic liquids, such as benzene, by mouth, students may be tempted to pipette water or salt solutions by mouth. However, it is impossible to guarantee the cleanliness of the pipettes or even the purity of the solutions. The first lesson in pipetting should include instruction in the use of pipette-filling bulbs.

2.8 SPLATTERING FROM HEATED TEST TUBES OR BEAKERS

One of the most common accidents in undergraduate laboratories is splattering of liquid from an overheated test tube or beaker. Since the substances splattered may include concentrated acids or bases or other corrosive materials, the possibility of this type of accident alone is sufficient to require that all persons in a laboratory wear safety glasses, even when not actually working. Unlike some of the other possibilities described in this chapter, splattering accidents are common and not easy to prevent in elementary laboratory courses. In addition to safety glasses, a lab coat, to protect clothing and perhaps prevent some burns, should be worn. It is important to avoid overcrowding in the laboratory as this can magnify the effect of splattering.

2.9 EYE AND FACE PROTECTION

Eyes are extremely vulnerable and *require* protection. Various types of glasses offer different levels of protection, and even ordinary prescription glasses without side panels are a great deal better than no glasses at all.

Contact lenses are not to be used in a laboratory. Not only do they offer no protection in themselves; they are unsafe even under safety goggles. Various fumes (for example, HCl) may be concentrated under the lenses and against the eye, with possibly serious results. Therefore, the use of contact lenses should not be permitted in the laboratory.

The various categories of eye protectors are described in the following paragraphs.

a / Ordinary Prescription Glasses

These offer some protection from chemical splash. However, they lack side panels so that this protection is quite limited; they may be adequate for some laboratories, such as those in which primarily physical measurements are being carried out and where there is limited danger of chemical splash.

Overall, eye protection requires safety lenses that are in compliance with the Standard for Occupational and Educational Eye and Face Protection (Z87.1–1968), established by the American National Standards Institute. This means impact resistance sufficient to withstand a 1-in. (2.5 cm) diameter steel ball dropped from a height of 50 ft (15 m), and a lens thickness of at least 3 mm. Ordinary prescription eyeglasses meet less stringent standards and are therefore not satisfactory for normal chemical laboratory use. (See NIOSH Publication No. 78–108, Tests of Plastic Plano Safety Spectacles)

b / Chemical Splash Protective Goggles

Several varieties of chemical splash protective goggles exist. These goggles fit over prescription lenses and either wrap around the face or have side panels to offer complete splash protection. They should have adequate ventilation (unless protection against vapor is also needed, in which case a tight seal against the face is required). The ventilation ports may be baffled, if splash protection requirements are severe. Some varieties are nonfogging and antistatic, to avoid clinging dust.

c / Face Shields
Complete coverage of the front of the face provides both splash protection and impact protection; this should obviously be used if
there is any possibility that the latter will be required. Face shields
should wrap around the face (essentially all commercial types do)
and, in addition, protect the top of the head and the neck as well
as the face. They should fit sufficiently comfortably to be worn for
as long as required.

d / Visitor's Goggles
Visitors are as much at risk as laboratory workers while they are in
the laboratory. Lightweight visitor's goggles are available commercially. Some form of eye protection must be worn by visitors while
in the lab.

e / Other Types of Eye Protection, Especially Against Light
Laser goggles are discussed briefly in Chapter 6, together with
lasers. There are special goggles to protect against ultraviolet light;
glass blowing goggles; and cobalt glass spectacles for looking into
furnaces, or into oxyhydrogen or oxyacetylene flames. In general,
these are specialized uses, and the appropriate type of protection is
well recognized and should be used.

In short, workers in locations in which chemicals are stored,
dispensed, or used, require eye protection. If there is any possibility that protection against impact for the entire face may be needed,
it must be provided as well.

2.10 EMERGENCY FACILITIES

a / Safety Showers
Every chemistry laboratory should be equipped with safety showers.
They are to be used immediately by laboratory workers who have
been splashed with toxic or corrosive chemicals (or whose clothing
has been) or whose clothing has caught fire. The shower should be
easily available. Proper housekeeping procedures ensure that access
to this and other safety equipment is not impeded. The shower,

once turned on, should deliver a copious stream of water, continuously and automatically, without the necessity for the affected individual to hold the shower on. This leaves both hands free to remove splashed clothing. The shower should be located near one of the laboratory exits and clearly designated by a sign. The floor should slope towards a floor drain. Emergency blankets should be located near showers, both to combat shock and to provide covering in place of removed clothing.

b / Eye Wash Station
With suitable eye protection, it is to be hoped that an eye wash station will not often have to be used. The principal remedy for acid, base, and practically all other spills of chemicals into the eye is to flush the eye *copiously* with water. An eye wash station should provide a soft stream or spray for an extended period—at an absolute minimum, 5 min. It should also be clearly marked and easily and rapidly accessible. Since more than the eye is likely to have been splashed (in fact, with proper goggles, it may be the rest of the face that is most in need of flushing), a full face spray is preferable.

c / First Aid and Burn Kits
At least one kit should be available within minutes, though not necessarily as rapidly as safety showers or eye wash stations. Various types of industrial kits are available. An aqueous paste of sodium bicarbonate is also useful as a temporary topical application for acid burns.

d / Emergency Blankets
These should be available near the safety shower, as mentioned previously. All items in this section (showers, eye wash stations, and first aid kits), should be checked regularly to be sure they are in working order, and available when needed.

2.11 PROTECTING GLASSWARE FROM BREAKAGE

With large containers of acids and bases, it may be worth covering the bottles with plastic netting. Also, padding may be used on the floor or table near where such bottles are stored or used. Plastic (for example, polycarbonate) ware may be used in place of glass for some items. Glassware should not be stored high on a shelf, where it would surely break if dropped, nor in such a location that the laboratory worker must reach across a crowded lab bench. Solutions that generate a large amount of heat upon mixing should not be prepared in soft glass bottles or cylinders. Glass desiccators, too, should not be exposed to elevated temperatures.

When carrying glassware, it is wise to be especially careful: use two hands; use a laboratory cart for more than one bottle; allow no running by anyone in a location where chemicals may be carried; and, in general, use common sense. Sponge off acid bottles, then carry them close to the body, so that the bottle will not be dropped in case of accidental jostling. Tongs or insulated grips must be used when glassware is hot; one cause of breakage and spills is attempting to pick up something too hot to handle.

2.12 HYGIENE

Chemists must be particularly careful to avoid carrying chemicals out of the laboratory on their hands or elsewhere on their body. Such simple precautions as washing with extra care before eating or using toilet facilities and showering after the day's work may reduce exposure if seriously toxic materials have been handled.

2.13 WARNING SIGNS AND LABELS

a / Warning Signs
Any areas with special hazards require warning signs. Standard signs exist for radioactive areas and radioactive hazards. These have been in use for over 30 years, and the radiation symbol is almost universally recognized. Specific information (X RAY, AIRBORNE RADIO-ACTIVITY, CONTAMINATION HAZARD), is also on the sign. Radiation signs are magenta on yellow.

Biological hazards also have a standard symbol, as does the "Laser Operating" sign. In addition, there are National Fire Protection Association (NFPA) standard symbols for labelling fire hazards (see Chapter 4). Such chemical or physical hazard warnings as "Explosive," "Oxidizer," "Flammable," "Poison," "Carcinogen," "High Voltage," are generally of the form of a standard Danger symbol, with the particular form of the danger stated underneath. Signs are normally red and black on white. *Caution* signs may be black and yellow. Areas with carcinogens should be labelled *"Danger—Presumptive Chemical Carcinogens—Authorized Personnel Only"*

Other signs are required to show the location of safety showers, eye wash stations, exits, and fire extinguishers. Fire extinguishers should show the type of fire (A, B, C) for which they are intended. See Table 4.4 for this classification of fires. Waste disposal should be labelled to show whether it is for glass, or organic waste, or other specific type. There may also be reminder signs (*No Smoking, Wear Goggles*). In general, the signs in a laboratory should be good enough that, in case of emergency, anyone in the laboratory including maintenance personnel and visitors, will know where to find appropriate equipment, exits, and so on. It is important to remark that it is possible to overdo the use of signs; a profusion of signs can be almost as confusing as no signs at all. Some judgment is therefore required. Suitable signs are commercially available at reasonable cost.

b / Labels

Chemicals require labels that state the hazards associated with their use. These should be on the containers themselves; also, the National Institute for Occupational Safety and Health (NIOSH) recommends that the entrance to an area in which there is likelihood of occupational exposure also be posted with a suitable warning. A typical example is the recommended warning for benzene:

Danger!
Benzene
Extremely Flammable
Keep away from heat, sparks, and open flame

Vapor Harmful
High concentrations of vapor
are hazardous to health
Provide adequate ventilation

"Occupational exposure" here means half the recommended limit.

(For benzene, the new limit is 1 ppm, time weighted average, or 5 ppm for any 15-min period.)

Too many signs, as previously mentioned, may be confusing; furthermore, too many signs probably indicates too many substances in the room atmosphere for safety. See Equation 5–2 of Chapter 5 for acceptable limits for different combinations of chemicals.

It is important to be sure the labels are read and properly understood. The area itself, if appropriate, should carry the carcinogen warning. (See also Chapter 3, Section 3.6 for specific information to be included on labels.)

2.14 SUMMARY

The precautions discussed in this chapter are for the most part sufficiently obvious that most practicing chemists are aware of them already. Whether all chemists actually follow the rules is another question. As discussed in Chapter 9, there are often excuses (not good reasons) for taking unsafe shortcuts. With the increased safety consciousness brought about by OSHA in the United States, it is likely that safe practice will more and more come to be expected. Students should be taught these principles in their first lab course.

3

Laboratory Equipment Hazards

Once the basic safety rules have been understood, the next step is to insure that the equipment in the laboratory is set up and operated properly. Unsafe equipment is an obvious path to disaster.

3.1 FLOWING WATER

In the laboratory, flowing water is generally used as the coolant in condensers or as a fluid for an aspirator for low vacuum operations. Such operations are usually continuous, sometimes over long periods, and it is therefore tempting to leave them unattended, often over-night. Various hazards are associated with unexpected variations in water pressure. A decrease in pressure may occur early in the working day, when water usage increases, or at unexpected times when there is a break or blockage in a water main. An increase in pressure typically occurs at the end of the day, when other water usage gen-erally decreases. Either event can cause trouble. Lowered pressure means slower flow, which can result in inadequate cooling of a dis-tillate and the escape of vapors. Increased pressure can cause leaks, can separate connections between tubing and condensers, can whip the end of the tubing out of the drain, or can even produce enough flow to exceed the capacity of the drain.

Among the various precautions, some are very easy to carry out and are adequate for *attended operations*. Others, more elaborate, are necessary for safe practice if the experiment is to be unattended, especially overnight.

a / Attended Operations

1. Make sure that all flexible hoses are free of defects such as cracks or slits, particularly at the ends.

2. Hoses should be assembled so as to avoid kinks that resist flow (Figure 3.1). They should be kept away from hot plates or flames that might soften or decompose them. Hose ends should be well into the receiving sink or drain.

3. All connections should be firm. For prolonged use, even when

Figure 3.1   Avoid kinks in rubber hose.

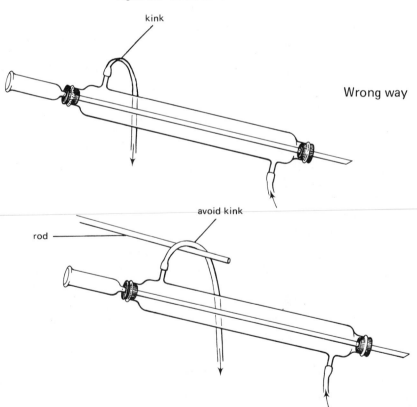

attended, they should be clamped or wired to valve ends and to condenser inlets and outlets. Be especially careful of tubing that has been well lubricated before being connected—it may slip off as easily as it went on.

4. A typically moderate water flow through a condenser is about 1 liter/min. Check the flow rate once or twice with a watch and a 1-liter beaker. Adjust the flow carefully at the start of the experiment by looking at the rate at which the water comes out of the exit tube. Do not try to estimate the flow rate by the turn of the water valve; this is completely unreliable. Finally, glance at the condenser from time to time during the experiment. A visible mist near the outlet signals incomplete condensation and suggests that water flow may have stopped.

b / Unattended Operations
Two categories of precaution are needed:

1. The water pressure should be regulated automatically. This objective can be realized by installing a regulator in the water main ahead of the valve that sets the desired flow. The regulator should, in turn, be protected by a suitable filter to prevent foreign particles, such as rust, from clogging it. (See part (i) below for sources of equipment.)

2. The water flow should be monitored so that, if it is interrupted, everything can be turned off—electrical heaters that power the experiment as well as the water supply itself. The latter is necessary because the water flow may have been interrupted by a break in a connection, which means that water can be spurting onto the floor. For this reason also, the monitor should be positioned at some point *after* the water has passed through the apparatus and is on its way to the drain.

c / Water-Flow Monitors
Various water flow monitors have been described in the literature. (See part (ii) below for references.) Conlon's modification, shown in Figure 3.2, is simple and reliable. When the water stops flowing, the water column drains through the lower capillary, the mercury column moves up on the right and down on the left, opening the

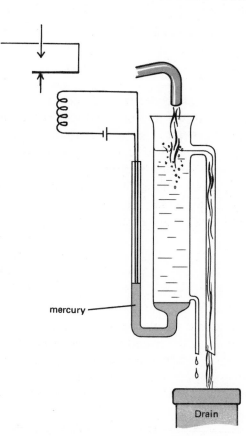

mercury

Figure 3.2 Cooling water-flow monitor. When the flow ceases, the column drains through the capillary and the mercury contact opens. The dimensions should be such that no mercury can drain through the capillary.

electrical contact and switch. If the monitor is connected to a solenoid valve in the water supply line, it can shut the water supply as well as the electrical power to the apparatus.

i / Sources of Equipment

Solenoid Valve, Catalog #826225. Automatic Switch Co., Florham Park N.J. 07932.

Water Pressure Regulator, Catalog #1/4-N263. Watts Regulator Co., P.O. Box 810, Lawrence, Mass. 01842.

Water Filter, Model #L1OU-3/4. Filterite Corp., Timonium, Md.

ii / Literature References to Water Flow Monitors

Burford, H. C.: "Improved Flow-Sensitive Switch," *J. Sci. Instr.,* 37:490(1960).

Conlon, D. R.: "Unattended Laboratory Operations," in *Safety in the Chemical Laboratory,* Division of Chemical Education, Am. Chem. Soc., Easton, Pa., 1967, pp. 83–86.

Cox, B. C.: "Protection of Diffusion Pumps Against Inadequate Cooling," *J. Sci. Instr.* 37:148 (1960).

Houghton, G.: "A Simple Water Failure Guard for Diffusion Pumps and Condensers," *J. Sci. Instr.* 33:199 (1956).

Mott, W. E., and Peters, C. J.: "Inexpensive Flow Switch for Laboratory Use," *Rev. Sci. Instr.* 32:1150 (1961).

Pike, E. R., and Price, D. A.: "Water Flow Controlled Electrical Switch," *Rev. Sci. Instr.* 30:1057 (1959).

Preston, J.: "Bellows Operated Water Switch." *J. Sci. Instr.* 36: 98 (1959).

3.2 COMPRESSED GAS CYLINDERS

Compressed gases are stored in steel cylinders, typically at an initial pressure about 2000 lb/in.^{2} (136 atm, or 138 bar) above normal atmospheric pressure, although sometimes even higher pressures are used. Condensible vapors, such as ammonia (NH_3), hydrogen chloride (HCl), or carbon dioxide (CO_2), are necessarily stored at their vapor pressures at room temperature.

These steel cylinders can be stored and handled safely, but careful practice is essential; if an accident does occur, it is likely to be a very serious one. We may classify safe practice into four categories: (a) identification; (b) storage and transfer; (c) dispensing of contents; (d) return of empty cylinders.

a / Identification

The U.S. Department of Transportation has established regulations for the labeling of compressed gas cylinders. Make sure you know the contents of any cylinder you use. Do not deface any label nor obscure it in use. Do not rely on labels affixed to caps (which are interchangeable) or rely on cylinder colors (which are not standardized).

If the labeling is unclear or defaced, as on an old cylinder, do not attempt to analyze the gas. Return the cylinder to the manufacturer.

b / Storage and Transport

A compressed gas cylinder whose head has been knocked off is a functioning rocket; the thrust can drive it through a masonry wall. Therefore, the valves must be protected and the cylinders must be secured to prevent them from falling when they are stored or moved (Figure 3.3). In the laboratory, cylinders must be strapped to the bench top or other firm support. The valve protection cap should be kept on until the cylinder is secured in place and ready for use. Cylinders should be moved only with a suitable hand truck, never rolled or dragged. They should not be subjected to temperatures above 50°C (122°F) and, of course, should never be subjected to a flame or be allowed to be part of an electrical circuit. Cylinders and valves are equipped with various safety devices such as a fusible plug that melts at about 70°C (158°F). Under no conditions should any of these be tampered with.

c / Dispensing of Contents

The outlet types and threads used for various categories of gases have been standardized to prevent the interchange of valves and gauges between gases that are incompatible. For instance, equipment that has been used on oil-pumped gases should not be used for oxygen, because residual internal oil films may then ignite. The standardization of thread types which prevents such misuse can be circumvented with the aid of various special adaptors, but such practices are very dangerous. In any case, make sure the threads are compatible; never force them.

Now refer to Figure 3.4 to help you follow the procedure for dispensing a compressed gas:

1. After the cylinder has been secured in place, remove the protecting cap and screw on the automatic pressure regulator. (Remember —watch the threads, some are right-handed, some left-handed.) Tighten the connection with a monkey wrench (*not* a pipe wrench, which roughens the edges of the nut).

Cylinder valves must be protected

Store properly

Secure cylinder before use

Don't drop

Transport correctly

Never tamper with safety devices in cylinders or valves

Keep gas out of breathing air

Return in condition received

Figure 3.3 Safe practice with compressed gas cylinders.

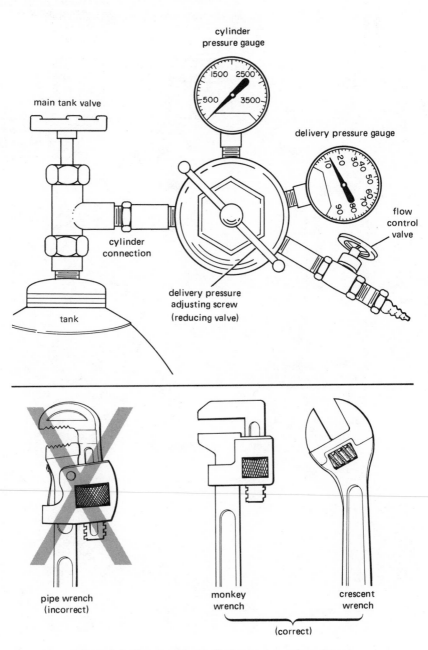

cylinder
pressure gauge

main tank valve

delivery pressure gauge

cylinder
connection

flow
control
valve

delivery pressure
adjusting screw
(reducing valve)

tank

pipe wrench
(incorrect)

monkey
wrench

crescent
wrench

(correct)

Figure 3.4   Compressed gas cylinder, gauges, and regulators.

2. Check that the flow control valve is closed, finger tight (it closes clockwise), as well as the delivery pressure screw (also called the reducing valve; it closes counterclockwise until it feels loose). Both gauges should read zero.
3. Now open the main tank valve fully. This may be hard, but it should be done by hand, not with a wrench. If necessary, use both hands. The cylinder pressure gauge should immediately show the full tank pressure. Next, turn the delivery pressure adjusting screw (clockwise) until the delivery pressure gauge indicates the pressure desired. The flow at this pressure may now be regulated by the flow control valve.
4. To shut off the gas supply, close the valves in the same sequence as they were opened. Shut the main tank valve first, and allow the residual compressed gas in the valves to drain out. When the gauges read zero, shut the delivery pressures screw (counterclockwise), and finally the flow control valve (finger tight).

d / Other Precautions

Do not use blasts of compressed gas for such purposes as blowing away dust—the rapidly moving particles can hurt.

Cylinders of toxic gases should be kept in such a manner that they cannot contaminate breathing air; this objective may require ventilation.

When corrosive gases are used, the cylinder valve should be worked frequently to prevent it from freezing. Regulators and valves should be removed and flushed with dry air or nitrogen after use, not left on the cylinder. When corrosive gases are to be discharged into a liquid, a trap should be provided to prevent any possibility of sucking the liquid back into the regulator or cylinder. Cylinders of corrosive gases such as hydrogen chloride may be protected by a drying tube when not in use.

If there is a fire, turn off the supply of gas *first,* then extinguish the fire. The reverse procedure could lead to buildup of combustible gas and explosion.

e / Return of Empty Cylinders

Do not empty a cylinder completely, because its residual contents may become contaminated if the valve is left open. For the same reason, do not store empty and full cylinders together. Never try to refill empty cylinders. Instead, remove the regulator, replace the cap, mark the cylinder "MT," and return it to the storeroom or supplier.

3.3 OTHER SYSTEMS CONTAINING COMPRESSED GASES

Chemists frequently have occasion to carry out experiments at elevated pressures. All such operations must be considered seriously hazardous and require extreme precautions.

First of all, the chemist should be aware of the possibility of inadvertent buildup of pressure. It is axiomatic that any operation to be carried out at atmospheric pressure should have some access to vent to the atmosphere. Sometimes the beginning student who is intrigued by the interchangeability of glass joints in a distillation setup forgets this. Mere venting, however, is not always enough, because other parts of a system may plug up unexpectedly, as by the crystallization of acetophenone (freezing point, 20°C) in a cold condenser. The chemist should be alert to all such risks.

Intentional high-pressure operations, especially in glass, should be treated by assuming that they will explode, although of course all reasonable preventive steps should be taken. Typical are hydrogenation with hydrogen gas, oxygenation with oxygen gas, or small sealed-tube reactions, sometimes run in an oven for many hours. Other reactions are run at a moderately elevated pressure to increase the solubility of a gas in the reaction mixture. In the latter instance, a pressure relief device should be used. A suitable one for a slight excess pressure is shown in Figure 3.5. Various commercial devices are also available for operation at higher pressures.

All glass apparatus under pressure should be shielded from people, preferably at three levels: First, if possible, the glass itself should be covered with wire mesh or with strong tape. Second, the equipment should be surrounded by an explosion shield. Finally, the chemist should be protected by a face mask.

When the experiment is over, the internal pressure in the system should be reduced to atmospheric pressure by a relief valve before the system is opened. If necessary, the system should first be cooled to room temperature.

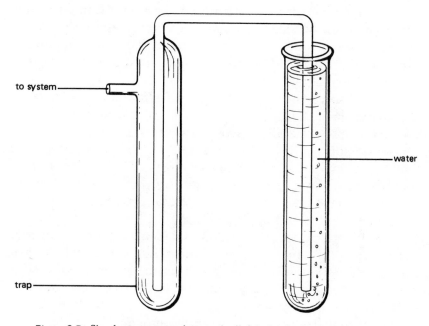

Figure 3.5   Simple pressure regulator and relief device for low excess pressures.

Note that at very high pressure (tens of kilobars and up), entirely new problems arise; laboratories carrying out work under such conditions are presumed to have designed the necessary heavy shields, and so on, for use in their work. The precautions recommended here are for a few atmospheres only.

3.4 VACUUM

Dangers associated with equipment under vacuum are, like those described in Section 3.3, related to a difference in pressure across the walls of a container. In an evacuated system the higher pressure is on the outside, not the inside, so that a break causes an implosion rather than an explosion. The resulting hazards consist of flying glass, spattered chemicals, and possibly fire.

A moderate vacuum, such as 10 torr (0.013 atm, or 0.015 bar), which can be achieved by a water aspirator, often seems safe compared to a "high" vacuum such as 10^{-5} torr (about 10^{-8} atm).

These numbers are deceptive, however, because the pressure differences between outside and inside are comparable (760 – 10, or 750 torr in the first instance, compared with $760 - 10^{-5}$, or 760 torr in the second instance). Therefore *any* evacuated container must be regarded as an implosion hazard.

Evacuated spaces that do not contain flammable or toxic chemicals, such as a vacuum desiccator or the vacuum space in a Dewar flask, can be provided with a satisfactory implosion safeguard in the form of wrapping with friction tape. For desiccators, a grid pattern of tape, which leaves the contents visible, will do. Dewar flasks should be taped completely, right up to the outside lip. Of course, the operator must also wear goggles. Various plastic desiccators now on the market reduce the implosion hazard.

Vacuum distillations often provide some of their own protection in the form of heating mantles, column insulation, and the like; however, this is not sufficient because an implosion would scatter hot, flammable liquid. An explosion shield and face mask should be used to protect the operator.

Equipment under vacuum is especially prone to rapid changes in pressure, which can establish large pressure differences within the apparatus. These pressures can push liquids into unwanted locations, sometimes with very unhappy consequences. There are three modes of protection against such transfers.

1. A nonregenerating trap such as shown in Figure 3.6(a), which is simply a reservoir that holds a given volume of liquid.
2. A regenerating trap shown in Figure 3.6(b) which holds a given volume of liquid but returns it to its source when the vacuum is restored. Assume that the "source of liquid" is the water aspirator, and orient the trap accordingly.
3. A check valve, which is a device that allows flow in only one direction. Water aspirators are normally provided with check valves that allow flow only from the system to the aspirator so that water cannot be sucked back into the apparatus if the aspirator should cease to function properly. However, these valves sometimes get clogged, and a trap should always be used between a water aspirator and the setup.

A final danger lies in the shock to mercury manometers that oc-

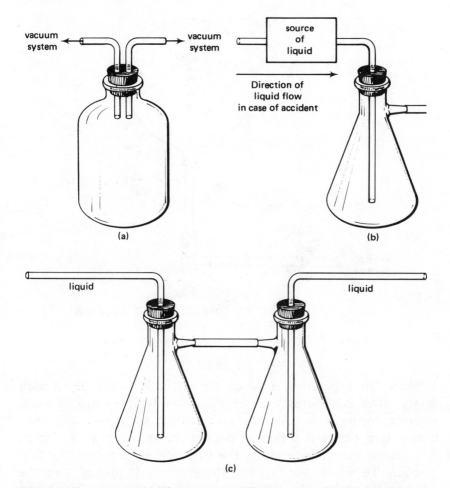

Figure 3.6 Vacuum traps. (a) When a pressure difference occurs, liquid is forced into trap and stays there. (b) Liquid is forced into trap by rise in pressure. When vacuum is restored, liquid is sucked back. The trap does not function in the reverse direction. (c) Double trap, positioned between two sources of liquid.

curs when a vacuum is released suddenly. The resulting surge can break the glass and scatter the mercury, a calamity that must be prevented. The best protection is to incorporate a capillary section into the manometer that will prevent any such surge even in the event of a sudden release of vacuum. Also, the manometer should be set in a container that will hold any spilled mercury. See Section 5.5 (q)(i), page 94.

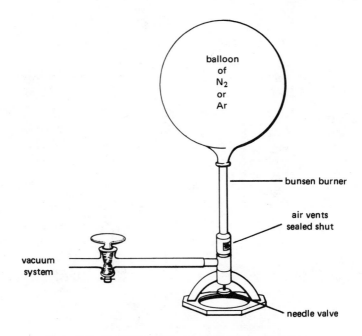

Figure 3.7 Simple system for controlled leak with inert gas.

When the experiment is over, the vacuum should be released slowly, after the contents of the flask have been allowed to attain ambient temperature. A very satisfactory and inexpensive controlled leak device is a Bunsen burner, set up as shown in Figure 3.7. During vacuum operation, the stopcock is turned to the shut position. To release the vacuum, the stopcock is opened, then the needle valve at the base of the burner is opened just enough to admit air slowly. If the system is to be inert, a balloon inflated with nitrogen or argon can first be tied to the mouth of the burner.

Vacuum pumps, just as other machinery with moving parts, require a guard for the belt.

3.5 ELECTRICAL HAZARDS

The presence of flammable liquids in chemistry laboratories aggravates the hazards that are normally associated with electrical equipment. One would certainly be alert to the danger of gasoline in a glass container stored near electrical appliances in the kitchen, yet many chemists think nothing of keeping a bottle of hexane, or even petroleum ether, near electrical stirrers, blowers, or switches. Furthermore, many experimental setups require special wiring, which is often carried out in a makeshift and therefore hazardous manner. The following recommendations, which are generally considered to be advisable for safe practice with electrical equipment, are therefore especially important for chemical laboratories:

a / Wiring Assemblies

Makeshift wiring assemblies, such as extension cords running the length of a workbench or, worse yet, bridging an aisle, must be avoided. Frayed cords or worn electrical insulation should never be tolerated. If a new outlet is needed, have an electrician install one with conduit or BX cable.

b / Grounding

All apparatus should be grounded. If your laboratory outlets do not accommodate the three-prong plug, use a "pigtail" adapter (Figure 3.8) that grounds through the screw on the cover plate.

c / Water

It should hardly be necessary to emphasize that your hands should be dry, and you should not be standing in water, when you handle

Figure 3.8 "Pigtail" adapter for electrical grounding.

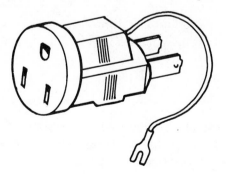

electrical equipment. Furthermore, if *any* liquid spills on a motor or other electrical device, disconnect the current and make sure everything (especially inside) is dry before reconnecting it. Also, additional care is needed with higher voltages. It is useful to work with one hand only, if possible. Metal watchbands form an electrode in a particularly sensitive location and must be avoided in any situation where the possibility of electrical shock exists.

d / Explosion Proofing
Ideally, all motors, switches, and other electrical equipment in a laboratory should be explosion proof, and all electrical receptacles should be equipped with spring-loaded cover plates to protect them when they are not in use. Certainly stirring motors for reaction flasks and other electrical equipment that can potentially be exposed to flammable vapors *must* be explosion proof. The same holds for refrigerators in which chemicals are stored. Remember, the inside of a refrigerator is a sealed space, and vapor concentrations even from slow leaks can therefore build up to hazardous levels. Ordinary refrigerators (but not the self-defrosting ones) can be made safe for laboratory use by removing the electrical controls, including lights, to the outside of the refrigerator.

e / Static Electricity
Finally, the laboratory worker must guard against the dangers of accumulation of static electricity, which may generate sparks. This is a recognized hazard in high voltage equipment, in which provisions for prevention of static buildups are usually incorporated into the design of the apparatus. (Nonetheless, the operator must be vigilant and aware of any malfunction.) More insidious are situations where static charge is unsuspected and comes as an unpleasant (or worse) surprise. For example, a chemist may clean out a piece of equipment with a flammable solvent such as acetone, then blow the vapor out with a stream of compressed air. The vapor pressure of acetone at room temperature is about 0.3 atm, and its concentration in air at equilibrium is therefore about 30 per cent by volume. Acetone vapor in air is explosive in the concentration range between 2.6 per cent and 13 per cent by volume. These values mean that as the chem-

ist blows the acetone out of the container, the mixture will pass entirely through the range of explosive concentrations. Any static buildup or friction spark is therefore dangerous, especially if the container is metallic and has only small openings as, for example, a steel gas-sampling tank.

Note that nonconductive surfaces moving past each other generate static electricity. Therefore, even certain forms of safety clothing, such as plastic shoe coverings, may be responsible for producing static electricity. It is advisable to use *grounded* conducting surfaces, and conducting materials, in cases where there is danger that explosive mixtures of gases may be present.

f / Electrical Connections

Connections, including ground connections, can degrade with age. An OSHA compliance program would require testing for continuity of conductors, for impedance of equipment ground, and for leakage currents (recall that currents of 10 milliamperes (mA) can be very serious, and in fact leakage currents must not exceed 1 mA). The Ecos Electronic Corp. claims that the instrument used by OSHA inspectors to test these points is their Model 1020.

3.6 LABELING AND STORAGE

There are two strongly recommended practices regarding the labeling of chemicals. The first is to make a habit of *reading* labels on bottles. Manufacturers' labels often include warnings of danger, precautions in use, content of impurities, and so on. Labels should also specify the concentrations of solutions—students should be trained to look for this information. The second practice, which is complementary to the first, is to label all chemicals informatively and durably, as described in the following paragraphs.

a / Information on a Label

As a *minimum,* a label on a stored chemical should indicate contents, source, date of acquisition, storage location, and brief indication of hazard. Of course, labels can tell more, such as a short summary of properties (melting point, boiling point), an indication of the degree of purity, and possibly a qualitative description of impurities. The importance of content labeling seems almost too obvious for comment, but the circumstances are not always so simple, as a scan of most storage shelves will attest.

Research samples, especially, are often deliberately labeled to be obscure to everyone but their owners, and even they may forget, after a few months, just what they wrote. Labels on research samples should therefore be keyed to a designated laboratory note book and page number. For chemicals of known identity, research groups should agree on a common naming system, such as that used in the *Merck Index,* or *Chemical Abstracts,* or perhaps the listing in a widely used chemical catalog. Some groups may prefer the nomenclature rules adopted by the International Union of Pure and Applied Chemistry (IUPAC). For use in student laboratories, the label on the bottle should correspond to that in the laboratory manual assigned to the class. Any instructor who has seen students pass up the bottle labelled 2-chloro-2-methylpropane while searching for *t*-butyl chloride will appreciate this need.

Chemicals obtained from supply houses bear their manufacturer's or distributor's names, but other laboratory samples often do not. It is helpful, when doubt arises about some particular sample, to know where to seek more information.

The date of acquisition is particularly important for chemicals that deteriorate with age and thus develop impurities, such as dioxane or other ethers that form peroxides (see Table 4.2).

Storage locations are hardly ever seen on labels, but they require little effort to note, and they help make laboratory operations easier as well as safer. Let us say that a laboratory or storeroom has a number of cabinets, shelf cases, refrigerators, and so on, where chemicals are stored. Each such cabinet or case has a number of shelves. The cabinets, or other storage areas should be numbered and the shelves lettered. Within a given shelf, no other designations are needed—the chemicals may be lined up alphabetically and, in any case, it is easy to find a chemical once you know what shelf it is on. Each chemical label then bears a (number + letter) designation, such as 5-C, which means that the chemical belongs on the C-shelf of cabinet number 5. Such labeling makes life easy because it practically ensures that chemicals will be returned to the places where they belong, and that they can be found there again. If the locations were originally chosen with the purpose of minimizing hazards, this good practice can be preserved.

b / Durability of Labels

It is prudent to assume that a stored chemical will remain on its shelf "permanently" (some seem to do so). A durable label therefore preserves the value of the chemical and is also a guide to safe disposal if the material is ultimately to be discarded. Ball point and soft-tip pen inks are often not permanent and should not be used. Pencil graphite does not fade, but smudges. "Permanent" blue-black ink is satisfactory, and so is the imprint from a typewriter ribbon. Best of all is india ink applied with an old-fashioned dip-in pen. Chemists used to cover their labels with melted paraffin; laboratory supply houses now offer label lacquers, and recipes for such coatings appear in the "laboratory arts" sections of chemical handbooks. See also Section 2.13 for other requirements for labels.

c / Safe Storage

Other aspects of safe storage are generally well known to chemists and technicians. Nonetheless, some cautions are worth emphasizing.

Chemicals should not be exposed to heat or direct sunlight. Most organic chemicals are light-sensitive, at least over long periods of time, and so are best kept in amber bottles in closed cabinets.

Organic liquids, and the sublimates from many solids, are good solvents and eventually cause the deterioration of many types of plastic caps and liners. However, Teflon is less permeable than most plastics, and Teflon cap liners, available from laboratory supply houses, are very helpful.

Metal screw caps, even if lined with paper or plastic, can eventually drop rust or other oxide particles into their bottles. Such particles may catalyze various decomposition reactions. In the case of ethers such as dioxane and isopropyl ether, metallic oxides catalyze the decomposition (possibly explosive) of any peroxide that may have been formed and allowed to accumulate.

It is good practice to survey chemical stocks regularly, perhaps annually, to cull out obsolete items, illegibly labeled containers, and any chemicals that show signs of decomposition. Some chemical supply houses will accept these materials for cash or credit.

4 Reaction Hazards, Flammable and Explosive Chemicals

4.1 INTRODUCTION

It is important to keep in mind some fundamental differences among the *rates* at which various hazardous reactions occur, because such considerations bear very strongly on the requirements for safe practice.

Ordinary open diffusion flames can propagate themselves at rates of the order of magnitude of a meter per second (m/sec), depending on the air-vapor concentrations in the burning mixture. Since the hand can move much faster than this, we find it easy to flick out a match flame by moving the vapor source (the hot match) faster than the flame can keep up with it. As another example, if you remove the mixing tube from a bunsen burner, you will find it difficult to keep a gas jet from the needle valve lit unless you restrict the gas flow to a very slow rate. Of course, gases react much faster at higher pressures, and flame propagations in an automobile cylinder can reach speeds of some 50 m/sec. True high explosives, however, are in another class. The shock wave in a piece of exploding TNT, for example, travels at a rate of about 5000 m/sec. Furthermore, the high explosive shock waves often are very directional in character, with results that can range from comic to deadly.

These considerations can be illustrated by a few examples that will serve as a background to recommendations for safe practice.

- An uncovered 400-ml beaker half-full of hexane catches fire. It burns leisurely, smokily. If left alone, it would burn itself out. It can be extinguished simply by setting a watch glass (use tongs!) over the beaker. It would be foolish to blast it at close range with a powerful jet of carbon dioxide, which could splatter its contents clear across the laboratory bench.

- A graduate student has finished distilling a mixture of hydrocarbon "over sodium" from a 2-liter flask fitted with a 24/40 standard-taper joint. He removes the flask, which contains a few milliliters of hydrocarbon residue, to clean it, and notes that some granules of sodium remain. That's hardly anything, he thinks, I'll just kill it with water. He holds the flask by its neck under the tap with his right hand, and with his left turns the cold water on full force. He hears a loud explosion, seemingly from some other part of the laboratory. As he turns back to the sink, he sees that he is holding a ground glass joint in his hand, and the sink is covered with powdered glass. The initial sodium-water spark had ignited the hydrocarbon-air mixture in the flask, parts of which were in the explosive range. But the force of the explosion was downward, and his hand is not even scratched. He has been lucky.

- A freshman student attempts to follow directions in a laboratory manual for an experiment on the "preparation and properties of oxygen." The gas is to be generated by heating potassium chlorate and is to be collected in bottles, in which small samples of phosphorus and sulfur are to be burned. The student becomes confused and places all three substances in his test tube, which he holds in his right hand. The contents flash, causing him to raise his left hand to protect his face. A fraction of a second later the contents detonate. Even though the burning had consumed all but a few hundred milligrams, the explosive force of the remaining mixture is enough to drive pieces of glass from the test tube through 50 sheets of his notebook. The student will lose his right eye, but his left eye has been saved by his upturned hand. Had he been wearing proper goggles, his right eye might have been saved as well. Of course, the three chemicals in his test tube were incompatible and should never have been set out together in a freshman laboratory.

4.2 ASSEMBLY OF APPARATUS FOR CHEMICAL REACTIONS

Most chemical reactions in the laboratory are carried out in glass equipment. Even the laboratory novice is always aware that glass is fragile and takes reasonable precautions most of the time. Nonetheless, the breakage of glass reaction equipment is a prime source of laboratory accidents, and extremely careful preventive action must be taken.

Since glass equipment is not inherently self-supporting, it must be linked to some other, immovable framework. The ultimate source of stability is, of course, the laboratory structure itself, particularly the bench top or the wall. A flat-based ringstand that sits on the bench top is suitable only for small-scale equipment. Remember that such stands can easily fall over backwards. Tripod-based stands are better. Any setup of substantial size (certainly any large enough to include a 1-liter reaction flask) should be supported by a framework that is rigidly attached to the work top or to the laboratory wall, or both. The setup should be as far back from the bench edge as possible. Thumbscrews on clamps should be tight. (Unless your grip is quite strong, the clamps should be finger-tight and then be given a 1/8 to 1/4 additional turn with a pair of pliers). If the setup is very high and heavy, such as a tall fractionating column, the usual 1/2-in. (1.25 cm) diameter rod is too flimsy as a main support. Instead, the main rod should be at least 3/4 in. (1.9 cm) in diameter and, of course, it should be securely fixed to a wall, not merely supported on the worktop.

The reaction equipment itself should be free from defects, internal stresses in the glass, and stresses produced by improper assembly. Flasks should first be examined carefully for "star" cracks, which are often seen at the very bottom of round flasks. Internal stresses are not obvious, but can be revealed with the aid of polarized light. Polarized-light kits are commercially available, or can be homemade from two sheets of polaroid film and a light source. For a large, costly, potentially hazardous reaction setup, such precautions are a good investment. To minimize stresses produced in the actual assembly of the equipment, the chemist must be alert to matters of alignment of connections, clamping, support of weights, and the like. The use of ball and socket joints and, where permissible, intervals of flexible tubing provide helpful safety factors.

4.3 FIRES

Two important properties of air-gas mixtures are significant determinants of their potential for starting a fire. These are (a) limits of flammability and (b) flash point.

a / Limits of Flammability

A hot platinum wire inserted into air containing 1 percent of methane by volume will catalyze the oxidation of any methane that comes in contact with it, but the reaction will not propagate away from the platinum surface. As a result, no fire or explosion will occur. The same wire inserted into a sample of methane containing 1 per cent of air will catalyze the reduction of any oxygen that comes in contact with it, but again, the reaction will not propagate and there will be no fire. The concentration limits within which such propagation *will* occur are about 5 per cent (lower flammable limit) and 14 per cent (upper flammable limit) of methane. These boundaries are also called lower and upper explosive limits, respectively, because a propagated flame is a branching-chain reaction that builds up pressure in a confined space, and very rapidly leads to explosion. Table 4.1 displays these limits for various gases and vapors.

b / Flash Point

It is common knowledge that an open beaker of diethyl ether set on the laboratory bench next to a bunsen flame will ignite, whereas a similar beaker of diethyl phthalate will not. The difference is that the ether has a much lower flash point. The flash point is the lowest temperature of the liquid at which its vaporization in air will reach the lower flammable limit. The flash point of diethyl ether, as determined in a closed container (the "closed cup" flash point) is −29°C, whereas for the phthalate (*p*-isomer) it is about 117°C. Therefore the open beaker of ether is dangerous near any source of ignition, whereas the phthalate becomes dangerous only when it is heated.

Flash points and flammable limits for various liquids are given in Table 4.1. A rough approximation for hydrocarbons may be obtained from the relationship:

$$\text{flash point, } ^\circ C \text{ (closed cup)} = 0.73t - 73 \qquad (4-1)$$

where t is the normal boiling point in degrees Celsius.

Table 4.1 PROPERTIES OF FLAMMABLE SUBSTANCES

Substance	Flammable Limits, % by Volume in Air		Flash Point (Closed Cup), °C
	Lower	Upper	
Amines			
methyl amine, CH_5N	5.0	21	gas
ethyl amine, C_2H_7N	3.5	14	–18°C
trimethyl amine, C_3H_9N	2.0	12	gas
Alcohols			
methyl alcohol, CH_4O	6.0	37	12
ethyl alcohol, C_2H_6O	3.3	19	13
iso propyl alcohol, C_3H_8O	2.3	13	12
2-pentanol, $C_5H_{12}O$	1.2	9.0	39
Aldehydes and Ketones			
formaldehyde, CH_2O	7.0	73	gas
acetaldehyde, C_2H_4O	4.0	55	–38
furfural, $C_5H_4O_2$	2.1		60
acetone, C_3H_6O	2.6	13	–20
methylethyl ketone, C_4H_8O	1.8	11	7
Esters			
methyl formate, $C_2H_4O_2$	5.9	20	–19
methyl acetate, $C_3H_6O_2$	3.2	16	–9
ethyl acetate, $C_4H_8O_2$	2.2	11	4
n-butyl acetate, $C_6H_{12}O_2$	1.4	7.6	27
p-diethyl phthalate, $C_{12}H_{14}O_2$			117
Ethers			
dimethyl ether, C_2H_6O	3.4	18	gas
diethyl ether, $C_4H_{10}O$	1.9	48	–45
ethylene oxide, C_2H_4O	3.0	80	<–18
dioxane, $C_4H_8O_2$	2.0	22	12

Table 4.1 (continued)

Substance	Flammable Limits, % by Volume in Air		Flash Point (Closed Cup), °C
	Lower	Upper	
Halides			
methyl chloride, CH_3Cl	8.1	17	gas
ethyl chloride, C_2H_5Cl	3.9	15	-50
methylene chloride, CH_2Cl_2	practically nonflammable		
n-butyl chloride, C_4H_9Cl	1.8	10	7
methyl bromide, CH_3Br	10	16	practically nonflammable
ethyl bromide, C_2H_5Br	6.8	11	
Hydrocarbons			
methane, CH_4	5.0	15	gas
ethane, C_2H_6	3.0	13	gas
propane, C_3H_8	2.3	9.4	gas
n-butane, C_4H_{10}	1.9	8.5	gas
n-pentane, C_5H_{12}	1.4	7.8	-49
n-hexane, C_6H_{14}	1.2	7.4	-30
n-heptane, C_7H_{16}	1.1	6.7	-4
n-octane, C_8H_{18}	1.0	4.7	13
benzene, C_6H_6	1.4	8.0	-17
toluene, C_7H_8	1.3	7.0	4
o-xylene, C_8H_{10}	1.0	6.0	32
Miscellaneous			
hydrogen, H_2	4.0	74	gas
ammonia, NH_3	16	25	gas
carbon monoxide, CO	12.5	74	gas
hydrogen sulfide, H_2S	4.3	46	gas
carbon disulfide, CS_2	1.3	44	-30

c / Mode of Ignition of Flammable Vapors

It does not take much to ignite flammable vapors. Any flame will do. So will a spark, however tiny. A catalytic surface, such as a bit of platinum-black, may also serve. Most combustible vapors are denser than air, which has an effective molecular weight of 29. Therefore, they settle down onto bench tops and floors, where they may accumulate and spread out horizontally and eventually reach an ignition source. Flammable vapors from massive sources such as spillages have also been known to descend into stairwells and elevator shafts and ignite on a lower story. If the path of vapor within the flammable range is continuous, the flame will propagate itself from the point of ignition back to its source.

The approach to safe practice follows logically from these principles. Flammable vapor mixtures and potential ignition sources should be both avoided unless they are an intrinsic part of an experiment and thus under complete control of the operator. The hazard of an open flame is obvious, but the surface of a hotplate, especially when heated at a "high" setting, can also ignite flammable vapors. Fibrous glass heating mantles are much safer. Oil baths, if not carefully-watched, can themselves ignite; silicone liquids are commercially available as nonflammable substitutes. For liquids that boil below 100°C, the safest source of heat is steam (generated without the use of an open flame).

Remember also that most flammable liquids are insoluble in and less dense than water. Thus, if they are poured down the drain, they may accumulate in traps or elsewhere as upper layers that generate flammable vapors. Disposal of such liquids is discussed in Chapter 7.

Another caution concerns atmospheres that are enriched in oxygen above the normal terrestrial concentration. The flammable limits in such mixtures are considerably broadened over those shown in Table 4.1. Even clothing exposed to an oxygen-enriched atmosphere becomes more highly flammable and can be ignited later by a hot cigarette ash.

The student should be alert to any unusual change in the appearance of a reaction mixture. Certainly the beginning of a rapid temperature rise or of fuming (such as the copious evolution of nitrogen dioxide from what was intended to be a nitration reaction) are signals for emergency protective measures, such as turning off heat sources, rapid application of a cooling bath, or leaving the scene.

A final point with regard to the danger of fire applies particularly to graduate students who have become familiar with a given reaction and feel confident about scaling it up to prepare a large quantity of product in a single run. Such a scale-up may also tempt the student to speed up the addition of reagent, say by adding 2 moles in the same time as he previously used for 0.2 mole.

The hazards in such scale-ups and speed-ups derive from the following circumstances. (a) Larger flasks offer a diminished surface/volume ratio for heat transfer; thus, overheating is a greater risk, especially when compounded with reaction speedup. (b) Thermal inertias are greater, and the observations of reaction behavior may be deceptive. For example, vigorous stirring of a 5-liter Grignard reaction mixture may cause the ether solvent to reflux so that the reaction appears to have started. Later, after the addition of more reagent, when the reaction really does start, the condensing system (which was the same as that previously used for a 1/2-liter flask) does not have the required capacity, and the reaction goes out of control.

4.4 EXPLOSIONS

It has been pointed out in the previous section that ignition of flammable material may initiate a branching chain reaction that rapidly attains explosive rates.

There are many substances, of course, that are explosive in themselves—sometimes in response to heat, sometimes to mechanical shock, and sometimes to contact with a catalyst. The decomposition of some explosives is autocatalytic, so that a bit of impurity can make trouble. A pure water-white sample of nitroglycerin in a cork-stoppered amber glass bottle might remain undisturbed in a laboratory storage cabinet for years (although this is not a recommended practice). However, a bit of nitric acid impurity would catalyze some decomposition, which would produce more nitric acid and release some heat, which could speed up matters still further, until. . . . In other instances a bit of highly sensitive explosive mixed with a larger quantity of insensitive but powerful explosive might start the trouble. Two glass containers of picric acid, for example, one pure and the other contaminated with an oxide of a metal such as lead, copper, or zinc, are inadvertently exposed to the same source of heat. The pure acid is rather insensitive, but not so the metallic picrates. Therefore, as a bit of the picrate salt is formed and is heated, it detonates, and the resulting shock sets off the unreacted picric acid.

There are various functional classes of substances that are typically explosive, such as peroxides, acetylides, ozonides, azides, and polynitrates. The research student must not be lulled by "familiarity" with any one of them. Any given sample may just be atypical enough (by virtue of impurities and so on, as noted above) to be especially dangerous. Furthermore, the hazard is associated not with the total energy released, but rather with the amazingly high *rate* of a detonation reaction. A high-order explosion of even milligram quantities can drive small fragments of glass or other matter deep into body tissues. The utmost in protection (explosion shield, goggles, and mask) are therefore essential when working with such substances.

a / Dust Explosions

To show future mine inspectors the power of a dust explosion, a few carts of coal dust are rolled into a mine while the "students" watch from high on an adjoining hillside. As the coal is set off with dynamite, a huge fireball shoots out of the mouth of the mine, the fragments of the carts flying out with it, and the hillside rumbles. Although dust explosions on such a scale are not likely in a laboratory, the chemist should nonetheless be aware that a suspension of oxidizable particles in air, such as magnesium powder, pollen grains, sawdust, or flowers of sulfur, constitute a powerful explosive mixture. The types of occasions in which such circumstances might arise include those in which a blast of air is used to blow away dust (a bad practice in any case), or in which an air stream is supporting a fluidized bed.

b / Explosive Gases

Special OSHA rules apply to hydrogen, acetylene, oxygen, and nitrous oxide. Any flammable or explosive compressed gas of course represents a special hazard. Hydrogen for use in buildings not in a special room, and exposed to other occupancies, is limited to 3000 ft^3 (85 m^3). OSHA regulations state that hydrogen systems shall be:

- In an adequately ventilated area (defined as follows: "Adequate ventilation to the outdoors shall be provided. Inlet openings shall be located near the floor in exterior walls only. Outlet openings shall be located at the high point of the room in exterior walls or roof. Inlet and outlet openings shall each have minimum total area of one square foot per 1000 cubic feet of room volume (0.09m^2 per 27 m^3). Discharge from outlet openings shall be directed or conducted to a safe location.")
- Twenty feet from stored flammable materials or oxidizing gases.
- Twenty-five feet from open flames, ordinary electrical equipment, or other sources of ignition.

- Twenty-five feet from concentrations of people.
- Fifty feet from intakes of ventilation or air conditioning equipment and air compressors.
- Fifty feet from other flammable gas storage.
- Protected against damage or injury due to falling objects or working activity in the area.

  More than one system of 3000 ft^3 or less may be installed in the same room, provided they are separated by at least 50 ft (15m). Each such system shall meet all of the requirements of this paragraph.

Although not all flammable gases require quite such stringent precautions, the example of hydrogen illustrates the type of precautions that are required. Requirements are slightly less stringent if a special room is set aside for the hydrogen system including the possibility of going to a maximum of 15,000 ft^3 (405m^3); a special building is required to go beyond this.

c / Liquefied Gases

In handling liquefied gases, remember that the substance is more concentrated than in the vapor phase, even if not under pressure, and that it may evaporate extremely rapidly. Oxygen in particular is an extreme hazard. Liquefied air is almost as bad, and if allowed to boil freely will increase in its percentage of oxygen (b.p. 90 K; –183°C) as the nitrogen (b.p. 77 K; –196°C) boils away first. If any liquefied gas is used in a closed system, pressure may build up, so that adequate venting is required. With flammable liquids, such as hydrogen, explosive concentrations may be achieved. In some cases it is difficult to translate OSHA rules directly into a laboratory situation, as any of the three problems, flammability, toxicity, and pressure buildup, may become serious.

Because of the relatively low heat capacity of liquefied gases, burns due to cold are rather easily avoided, unless massive spills are involved. Liquid nitrogen boiling freely in an open small Dewar flask (properly taped) is rarely any type of hazard, unless one puts one's fingers into the flask.

4.5 STORAGE OF SOLVENTS AND OTHER FLAMMABLE COMPOUNDS

There are two major aspects of safe practice related to the storage of flammable solvents: (a) limit the quantity to be stored and the time of storage; (b) use proper storage equipment.

a / Quantities and Storage Life

There is no reason for any laboratory to store a year's supply of solvents. At the other extreme, no research group would take kindly to a schedule of daily replacement of small volumes of solvents from the storeroom. Some reasonable compromise should be based on the frequency with which flammable liquids are needed, the quantities used, and the types of containers available. For example, it is usually unreasonable for a research laboratory to store more than a gallon of any flammable liquid; moreover, such quantities should always be kept in safety cans [see part (b) of this section and Table 4.3]. Furthermore, any such material that is to be kept "permanently" for occasional use should not require large quantities—perhaps a half liter at most.

For student laboratory classes, solvents and reagents should be set out only for the time during which a particular experiment is scheduled (typically one week), then removed and replaced.

Particular attention must be given to compounds that may form peroxides and hydroperoxides. Most chemists think of ethers when peroxidizable compounds are mentioned, but the class is more inclusive. In general, the ease of peroxidation is associated with organic structures of the general type.

$$
\begin{array}{c}
| \\
-\,C\,-\,Z\,- \\
| \\
H
\end{array}
$$

where Z is the electron-rich group such as oxy ($-\ddot{\mathrm{O}}\,-$), vinylene ($-\,CH = CH\,-$), amine ($-\,\dot{\mathrm{N}}\,-$), and so on, and where H is preferably a tertiary or secondary hydrogen rather than a primary one. The worst offenders are ethers, acetals, dienes, vinylacetylenes, vinyl monomers (including acrylates), amides, and various olefins with allylic hydrogen, chlorine, or fluorine substituents. Even aromatic hydrocarbons with tertiary benzylic hydrogens, such as cumene, must be included, as must secondary alcohols. Various inorganics are also peroxidizable, including alkali metals (especially potassium) and their alkoxides and amides, and many other organometallics.

The most dangerous of these substances react with air to produce explosive peroxides even without concentration. Volatility aggravates the problem because small quantities may evaporate into the cap area, where even very slight leaks make some air accessible and result in an accumulation of peroxide in the most hazardous area of all—around the threads of the cap where a twist is just enough to detonate it. Recall the previous warnings about the destructive potential of even small quantities high explosive. If this is not bad enough, the dispersed liquid also ignites and spreads flame around the area. These are no idle speculations. Old bottles of isopropyl ether have killed in just this way. In one instance, an unused hood, upon being cleaned out for laboratory renovation, was found to contain a 20-year-old gallon bottle half-full of dioxane. Some white crystalline material had accumulated around a rusty metal cap. The building was evacuated while a professional bomb squad removed the bottle to an isolated firing range, where the impact of a rifle bullet set off a high-order detonation. The procedure was costly and disruptive, but no one was hurt. Table 4.2 lists typical peroxidizable compounds in different categories of hazard, and suggests appropriate labeling procedures.

i / Detection of Peroxides

The method recommended by Jackson and coworkers follows.*

Reagent: Add about 100 mg sodium iodide (NaI) or potassium iodide (KI) crystals to 1.0 mL of glacial acetic acid.

Procedure: Add about 0.5 to 1.0 mL of the material being tested to an equal volume of the reagent. A yellow color indicates a low concentration ($\sim$ 0.01 per cent) and brown a high concentration of peroxide in the sample. A blank should be run, using some nonperoxidizable compound such as pure *n*-hexane.

*J. Chem. Educ., 47:A175 (1970)

Table 4.2 EXAMPLES OF PEROXIDIZABLE COMPOUNDS

RED LABEL—PEROXIDE HAZARD ON STORAGE
DISCARD AFTER THREE MONTHS

isopropyl ether
divinyl acetylene
vinylidene chloride
potassium metal
sodium amide

YELLOW LABEL—PEROXIDE HAZARD ON CONCENTRATION
DISCARD AFTER ONE YEAR

diethyl ether	dicyclopentadiene
tetrahydrofuran	diacetylene
dioxane	methyl acetylene
acetal	cumene
methyl isobutyl ketone	tetrahydronaphthalene (Tetralin)
ethylene glycol dimethyl ether (glyme)	cyclohexene
vinyl ethers	methylcyclopentane

YELLOW LABEL—HAZARDOUS DUE TO PEROXIDE
INITIATION OF POLYMERIZATION*
DISCARD AFTER ONE YEAR

methyl methacrylate	chlorotrifluoroethylene
styrene	vinyl acetylene
acrylic acid	vinyl acetate
acrylonitrile	vinyl chloride
butadiene	vinyl pyridine
tetrafluoroethylene	chloroprene

*Under conditions of storage in the liquid state the peroxide-forming potential increases and certain of these monomers (especially butadiene, chloroprene, and tetrafluoroethylene) should then be considered as A list compounds.

From H. L. Jackson, W. B. McCormack, C. S. Rondestvedt, K. C. Smeltz, and I. E. Viele: "Safety in the Chemical Laboratory. LXI: Control of Peroxidizable Compounds." *J. Chem. Educ.* 47(3): A176 (March, 1970).

ii / Removal of Peroxides

There are two recommended methods.

1. Pass the solvent through a short column of activated alumina. No water is thereby introduced. The alumina catalyzes the decomposition of many peroxides, but it is possible that some peroxides may be retained unchanged on the column. The alumina should therefore be disposed of in a manner appropriate for a flammable material.

2. Make up a reducing solution from 60 g ferrous sulfate ($FeSO_4$), 6 mL concentrated sulfuric acid (H_2SO_4), and 110 mL water. Shake

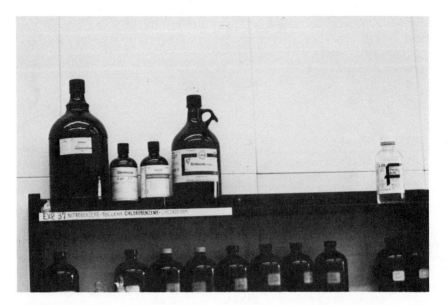

Figure 4.1 Poor storage. Chemicals for a student experiment are set out neatly, but in a very dangerous manner. Note the very toxic and flammable solvents in large glass bottles on an open, high shelf, with an extremely powerful oxidizing agent, perchloric acid, nearby. Note also that the highly toxic benzene is being used instead of the less toxic toluene. The bottles, from the left, are chloroform, chlorobenzene, benzene, and nitrobenzene (the latter is itself extremely dangerous).

the sample with this solution to remove the peroxide. Of course, the resulting sample is wet and may have to be dried with a suitable desiccant.

b / Storage Equipment

A suitable container for a flammable liquid must not break when dropped, its cap must be self-closing, and it must be provided with a pressure-releasing mechanism that will prevent release of all its contents when it is exposed to heat. These features are embodied in "safety" cans, which are available from various laboratory supply houses and other commercial sources in sizes from 1 pt (0.5 L) through 5 gal (19 L) and in various shapes. These cans offer important, indeed overriding, safety advantages, including those cited above. Perhaps most significant is the fact that, if a fire starts nearby, the internal pressure resulting from the outside heat causes the spring-loaded cap to release some vapor from time to time, but the bulk of the liquid is retained. This behavior may prevent a small fire from turning into a disaster and makes it much safer for fire fighters to operate on the premises. Safety cans should also be used for disposal of flammable liquid wastes in student laboratories. An open container for such use is obviously unsuitable, and the usual screw-capped bottle does not allow for pressure relief when incompatible liquids are poured in.

When flammable liquids, even in safety cans, are not in use, they should be stored in metal cabinets, preferably of the double-walled variety specified by the National Fire Prevention Association. Some storage practices that could lead to dangerous accidents are illustrated in Figure 4.1.

Storage rules promulgated by OSHA are summarized in Table 4.3.

4.6 LEAKAGES OF VAPORS DURING TRANSFER OPERATIONS AND CLEANUPS

Vapor is released to the atmosphere by displacement from a container that is being filled with a volatile liquid, and by evaporation from drippings. Therefore, all transfers of volatile liquids should be conducted in the hood. If drippings are sufficiently volatile, they may be left in the hood until they evaporate. Otherwise they should be wiped (include the shoulder of the reagent bottle, if necessary), and the wiping cloth or paper towel should be placed in a closed container for disposal.

Likewise, if a stream of compressed gas used in a chemical reaction is not completely consumed, the unreacted portion is released into the laboratory air, unless it is scavenged by a suitable trap. The experimenter will generally be aware of appropriate liquid reagents that can be used in bubblers. (Don't forget to add a safety trap as shown in Figure 3.6.) Less familiar is a family of granular solids that are usually more convenient and sometimes even more efficient than liquid bubblers. A good compromise among granule size, trapping efficiency, and avoidance of excess resistance to gas flow can be achieved by choosing granules somewhere within the range of 6- to 20-mesh size (particle diameters about 1 to 3 mm), sandwiched between thin layers of glass wool, and deep enough in

Table 4.3 MAXIMUM ALLOWABLE SIZE OF VARIOUS CONTAINERS FOR FLAMMABLE AND COMBUSTIBLE LIQUIDS

Liquid Classification	Glass or Approved Plastic	Metal (Other Than DOT* Drums)	Safety Cans	Metal Drums (DOT* Specifications)
Class 1A (Flash point below 22.8°C. Boiling point below 37.8°C)	1 pint†	1 gallon	2 gallons	60 gallons
Class 1B (Flash point below 22.8°C. Boiling point at or above 37.8°C)	1 quart†	5 gallons	5 gallons	60 gallons
Class 1C (Flash point at or above 22.8°C. Boiling point below 37.8°C)	1 gallon	5 gallons	5 gallons	60 gallons
Class II (Flash point at or above 37.8°C and below 60°C)	"	"	"	"
Class III (Flash point at or above 60°C and below 93.3°C)	"	"	"	"

*DOT—U.S. Department of Transportation
†1 gallon is allowed if the substance cannot be kept in metal or if the procedure requires more.
1 pint = 473 mL; 1 quart = 946 mL; 1 gallon = 3.8 liters.
Design, Construction, and Capacity of Storage Cabinets for Flammable and Combustible Liquids (Based on OSHA 2077, 1910.106 (d) 3)

1. *Maximum Capacity*—Not more than 60 gal of Class I or Class II liquids, nor more than 120 gal of Class III liquids may be stored in a storage cabinet.
2. *Fire Resistance*—Storage cabinets shall be designed and constructed to limit the internal temperature to not more than 325°F (170°C) when subjected to a 10-min fire test using the standard time-temperature curve as set forth in *Standard Methods of Fire Tests of Building Construction and Materials* (NFPA–251–1969). All joints and seams shall remain tight and the door shall remain securely closed during the fire test. Cabinets shall be labelled in conspicuous lettering "Flammable—Keep Fire Away."
   (a) Metal cabinets constructed in the following manner shall be deemed to be in compliance. The bottom, top, door, and sides of the cabinet shall be at least No. 18 gauge sheet iron and double walled with 1½-in. air space. Joints shall be riveted, welded, or made tight by some equally effective means. The door shall be provided with a three-point lock, and the door sill shall be raised at least 2-in. (5 cm.) above the bottom of the cabinet.
   (b) Wooden cabinets constructed in the following manner shall be deemed in compliance. The bottom, sides, and top shall be constructed of an approved grade of plywood at least 1 in. (2.5 cm.) in thickness, which shall not break down or delaminate under fire conditions. All joints shall be rabbetted and shall be fastened in two directions with flat-head wood screws. When more than one door is used, there shall be a rabbetted overlap of not less than 1 in. (2.5 cm.). Hinges shall be mounted in such a manner as not to lose their holding capacity as a result of loosening or burning out of the screws when subjected to the fire test.

the direction of gas flow so that a reaction time of about half a second is provided for as long as the experiment continues. For some small-scale applications, the familiar "drying tube" will suffice. For others a larger capacity will be needed. Some granular reagents useful for trapping vapors are soda-lime, for vapors of acids; activated alumina impregnated with potassium permangante ($KMnO_4$) and then dried, for formaldehyde and other readily oxidizable vapors of low molecular weight; and granular activated carbon, for vapors of bromine, iodine, and odorous vapors in general.

a / Dismantling and Cleaning Equipment

These operations can be a significant source of room contamination. Washing stenched apparatus in the sink under running hot water negates all previous care that might have been exercised in vapor control. The following procedure should be used: Provide a pot half-filled with granular activated carbon of the type used for air purification. When a piece of apparatus that needs decontamination is emptied or removed from a set-up, it should first be inverted into the pot of carbon and left there until all the vapor has been adsorbed. (For example, after use with liquid bromine, an empty graduated cylinder contains residual bromine drippings and vapor. It is inverted into the pot of carbon. Within a minute, all the bromine vapor will have been adsorbed and the removed cylinder has no bromine odor and can be washed and replaced in the cupboard.) Carbon used in this way has a capacity up to 1/3 of its own weight and will usually last for weeks or months. When it is no longer effective, it may be put in a plastic bag and dumped into the trash can.

b / Storage of Bottles or Ampules of Stenches and Fuming Liquids

Liquids such as bromine, chlorosulfonic acid, and fuming nitric (HNO_3) and sulfuric acid (H_2SO_4) solutions discharge vapor by leakage out of glass stoppered bottles and especially out of ampules that have been opened and resealed with cork. Such leakages accelerate the corrosion of metalware and instruments in the laboratory. When such containers are stored in the hood, they take up valuable work space and progressively attack the hood ducts. Stenches, such as mercaptans, although not corrosive, are offensive. In the case of ampules, it is best to reseal them after each use. One suitable device for storage of stenches is the "static hood," available from Bel-Art Products, Pequannock, New Jersey 07440. Another possibility is an old, inoperative refrigerator containing open dishes of granular activated carbon.

4.7 EMERGENCY PROCEDURES

Among the most common emergencies are small bench-top fires, which are normally extinguished without summoning the fire department or evacuating the laboratory. Nonetheless, there is always some danger that a fire may spread, and therefore the person in charge should be alert to the possible need for emergency procedures if matters should get out of hand. The transition from a trivial fire (or other emergency) to a major problem, when it occurs, may be very rapid. In such cases, the following set of actions is required:

1. *Alert personnel in the vicinity.* Every one in the laboratory should be immediately informed of the nature and extent of the problem and of the action expected of them. Personnel in nearby laboratories should also be alerted.
2. *Confine the Emergency.* Such confinement may be as simple as pulling down the sash to a hood, or it may require shutting the door to the laboratory (or transoms between laboratories) or making sure that all fire doors in the vicinity are shut, as they should normally be.
3. *Evacuate the building.* This works well only where there has been a program of periodic drills. Never feel guilty that you may have "overreacted." You must realize that a successful practitioner of safe procedures will be characterized as one who makes a fuss, or is a nuisance, even though no catastrophes have occurred. (Of course, it is the "fuss" that prevents catastrophes.)
4. *Summon aid.* The prime resource is your local fire department. Of course, supply all relevant information about the emergency —what chemicals or compressed gases are present, in what quantities, and so on.

The actions to control a "trivial" fire or other emergency are usually taken by the person in charge of the laboratory, or by the operator carrying out the experiment involved. Professional fire fighters are needed when conditions are more severe. The division between the two categories is not always sharp, however.

The first step, should it be needed, is the rescue of injured or trapped occupants. If the laboratory atmosphere is contaminated, rescue personnel must be provided with self-contained air masks that have at least a 15-min capacity. If a fire is out of control, the rescuer must wear an aluminized heat-reflecting suit. Of course, such capabilities are available only when the equipment is readily at hand and personnel have been trained in its use.

The control of small chemical fires is usually carried out with carbon dioxide extinguishers, which are provided as the basic equipment in most laboratories. For fires that cannot be reached or controlled by carbon dioxide, the dry chemical extinguishers (sodium or potassium bicarbonate) are much better. Water-type extinguishers are suitable for trash fires in waste baskets and the like. The standard classification of fires is shown in Table 4.4.

If chemicals that require special extinguishers are in use, the laboratory personnel should familiarize themselves with the needs and provide the necessary equipment.

Table 4.4 CLASSIFICATION OF FIRES

Class A Fires: These involve ordinary combustible solids such as paper, wood, coal, rubber, and textiles. They are accompanied by destructive distillation, producing vapors and flame and leaving glowing embers. They can be extinguished by quenching with water.

Class B Fires: These involve petroleum hydrocarbons such as diesel fuel, motor oil, or greases, as well as volatile combustible liquids such as the various flammable solvents. To extinguish these fires, air must be excluded by blanketing with a foam, inert gas, or with volatile halogenated hydrocarbons.

Class C Fires: These involve electrical equipment and the danger of electric shock. If the current is turned off, they may be treated as Class A fires; otherwise agents suited for Class B fires must be used.

Class D Fires: These involve combustible and reactive metals, such as sodium, potassium, lithium, magnesium, zirconium, and titanium, and their various alloys, hydrides, and organometallic compounds. They must be extinguished with dry nonreactive powders (*not* bicarbonate-based powders). These approved powders are generally proprietary formulations, which may include combinations of talc, sand, alkali metal salts, graphite, and the like.

Many buildings have fire hoses as part of their original equipment. If the nozzles are of the customary straight-tube type, they are unsuitable for laboratory fires because they produce a hard stream that can break bottles and make matters worse. Tube nozzles should be replaced with spray-type nozzles equipped with shut off valves.

Finally, fire extinguishers should be labeled according to the types of fires for which they may be used.

Specific standards, derived from National Fire Protection Association standards, apply to sprinkler systems and certain other aspects of fire protection. (See Section L, § 156–165, of OSHA 2077).

5

Toxic and Corrosive Substances

5.1 MORE RULES

Practically all of the chemicals in the laboratory are toxic, and some are corrosive as well. It is dangerous to permit routine spillage or exposure either to solvents or to vapors. Furthermore, since traces of highly toxic material may settle on almost anything in the laboratory, food and drink must not be brought into an area in which chemicals are stored or handled. In fact, most of the basic laboratory rules discussed in Chapter 2 apply at least in part because of the possibility of contamination of workers, food, clothing or other articles in the laboratory.

In addition to the prohibition against eating, drinking, or smoking in the laboratory, or testing chemicals by taste or odor, the most important rules pertaining to toxic and corrosive chemicals are:

1. Do not pipette by mouth.
2. Provide such safety equipment as showers and eye wash stations.
3. Maintain good housekeeping procedures.
4. Avoid horseplay.

As with all rules, it is well to be sure that they are understood by all personnel (and students), and that the reasons for them are understood. Several points need to be considered in explanations of the rules; the most important are the relative importance of acute and chronic toxicity, and the routes of entry of the various substances that may be encountered.

5.2 ACUTE AND CHRONIC TOXICITY

Acute poisons produce their effects in a relatively short time, generally from a single concentrated dose of the toxin. Chronic effects, however, are the result of low doses, built up over time. All carcinogens, for example, are chronic poisons. Generally, an incident of acute poisoning will result in one of three outcomes: death, permanent disability, or recovery. Carbon monoxide poisoning is a good example. If one is not killed by the gas, one may suffer brain damage. Assuming neither death nor irreversible brain damage results, one generally expects essentially complete recovery. However, as with many acute poisons, carbon monoxide may also be a chronic poison at much lower doses. There is evidence that those already suffering from heart disease, angina, and other physical ailments may be seriously affected by exposure to carbon monoxide levels found in rush-hour traffic. At one time the most probable cause of acute poisoning in the chemistry laboratory was hydrogen sulfide (H_2S). With the switch to thioacetamide, this hazard is far less common. Acute poisoning by nitrogen dioxide (NO_2) formed by reduction of nitric acid occurs occasionally.

A large number of *chronic* poisons are handled regularly by chemists. Some of the poisons are well known, and have been for some time; the "Mad Hatter" suffered from mercury poisoning, and hatters typically "went mad" because of the use of mercuric nitrate [$Hg(NO_3)_2$] in sizing felt. Chronic toxins are insidious, wreaking their damage over years; sometimes the symptoms, as of certain respiratory diseases (chronic bronchitis, emphysema) are not sufficiently specific to tie the disease directly to the cause, such as inhaling chlorine gas (Cl_2) at low concentrations over many years. Most cancers fall into the same category. There are exceptions; asbestosis, a lung disease caused by asbestos, is unmistakable, as is mesothelioma, an asbestos-caused cancer that typically takes 30 years to develop after initial exposure. Asbestos is one of the chronic toxins that are cumulative; the body does not eliminate them or metabolize them significantly so that repeated or continual low doses may build up to toxic levels. The effects of some heavy metals follow a similar pattern, as do, for example, certain organochlorine compounds, like the pesticide, kepone.

a / Routes of Entry

There are three main routes of entry to the body for toxins: inhalation, ingestion, or passage through the skin. Obviously, vapors normally enter by inhalation; vapors form a large class, ranging from mercury to hydrogen sulfide to solvent vapors, among others. Once breathed in, they may irritate the upper respiratory tract; soluble irritants, such as hydrogen chloride (HC1) or sulfur dioxide (SO_2), typically do this. Insoluble vapors such as nitric oxide (NO) or carbon monoxide (CO) are not dissolved by the bronchi. They penetrate to the alveoli, or air sacs, of the lung, and, having reached the alveoli, may enter the blood stream in the same manner as oxygen. Some gases (chlorine or hydrogen sulfide, for instance) affect the entire respiratory tract. Since vapors are often particularly difficult to deal with in terms of laboratory safety, we will have to keep inhalation clearly in mind as an important route of entry.

The second route of entry, through the mouth, can be eliminated almost entirely if proper laboratory procedures (no food in lab, no pipetting by mouth, and so on) are followed. Otherwise, the results may be deadly.

The skin may be a more serious route of entry than it is normally considered to be. Anyone who has taken a course in qualitative analysis probably knows that the skin itself can be damaged; the yellow color resulting from the denaturing of the skin proteins by nitric acid, and the black color from contact with silver nitrate ($AgNO_3$) are all too familiar, although apparently not too serious in themselves. Furthermore, it does not come as a major surprise to learn that solvents, such as acetone, can dissolve skin oils, causing the skin to dry and crack. However, it is most important to note that many solvents, such as benzene, phenol, and the nitroanilines, to pick a few horrible examples, can pass directly through the skin. In addition, these solvents, and others, may act as vehicles, carrying a variety of other substances through the skin as well.

A final point is worth noting: many chemicals thought at one time to be innocuous have been found to be very dangerous, often because they are carcinogenic. For example, vinyl chloride was at one time considered a promising anesthetic and is a common industrial chemical. Its best known current use is the manufacture of poly-vinyl chloride, one of our most common plastics. We now know that vinyl chloride is a potent carcinogen, causing angiosarcoma of the liver; this relatively rare cancer is at present always fatal. Permissible occupational exposure to vinyl chloride was rather abruptly lowered 500-fold, and a drop to zero exposure seems advisable. See Section 5.5 (a) (vi), page 83.

Other chemicals that are much more common in the laboratory are now treated with vastly more respect. Benzene and carbon tetrachloride can no longer be used freely or carelessly. Both cause serious liver damage; in addition, benzene can cause kidney damage and leukemia, and carbon tetrachloride may be carcinogenic. It is now known that these substances should be handled with extreme care, only in the hood, and with protective clothing. If possible, other substances should be substituted for these, as we shall discuss later.

Among corrosive chemicals, strong acids and bases, as well as liquids that are strong dehydrating or oxidizing agents, eat away the skin. In addition, all corrosive chemicals are threats to eyes, where of course the damage becomes permanent. With this as general background, we can now begin to consider specific problems.

5.3 CORROSIVE CHEMICALS

a / Strong Acids

It is well known that strong acids, especially if concentrated, can attack the skin and permanently damage the eyes. In addition, specific acids have particular properties, depending on whether they are also oxidizing agents, dehydrating agents, and so on. An acid spill is generally not, however, an instant disaster; it takes at least some seconds to begin to burn the skin, which should allow sufficient time to reach water. Some acids, such as hydrochloric, take a relatively long time to act. Even sulfuric and acetic acids, both dehydrating agents, can be washed off if one acts with sufficient alacrity. An exception is chromic acid, which exists as a mixture with sulfuric acid prepared as "cleaning solution." It is a strong oxidizing agent which acts extremely rapidly. Nitric acid (HNO_3), as mentioned earlier, denatures skin protein, and its vapors are extremely toxic. Hydrofluoric acid (HF) is extraordinarily corrosive to skin and mucous membranes, producing burns which are very slow to heal. Hydrofluoric acid is not suitable for use in student laboratories at all, in fact. If spilled on the skin near the fingernails, the result is not merely painful; it may cause loss of the finger, as the burn works its way to the bone.

It should be obvious that one does not pour a strong base onto spilled acid; even after students have learned titrations, they should not expect to come close to exact neutrality. Nevertheless, students in high school and freshman laboratories have been known to try to neutralize acid spilled on the skin with strong base, and it is therefore incumbent on the instructor to warn the students before the first laboratory period in which they will be handling chemicals that they should not fall into this particular trap, which turns a mere accident into a bonafide disaster.

b / Strong Bases

The most common are sodium (NaOH) and potassium hydroxide (KOH), calcium hydroxide [$Ca(OH)_2$], and ammonia (NH_3). The metal hydroxides, especially the alkali metal hydroxides, are extremely destructive both to skin and to the tissue of the eyes. If touched in the form of dry pellets, there is time to wash off the chemical with *copious* quantities of water. A little water is much worse than none: it should be remembered that the skin provides enough water for the bases to begin to dissolve, and thus attack the skin. These bases should not be added in solid form to a reaction mixture when a concentrated solution is called for, as the large heat of solution will raise the temperature to a possibly dangerous level.

Ammonia is a strong irritant, and quite soluble in water. Therefore, assuming one does not inhale a large enough dose to produce death by bronchial spasm, it is unlikely that one will be able to stay in the vicinity of the gas for long enough to be seriously injured. It is such a strong irritant that it will normally be used in the hood in any case.

c / Vapors

Vapors (or aerosols, which behave much like vapors for practical purposes) of sulfuric, phosphoric, and hydrochloric acids and of other concentrated acids are extremely irritating to the bronchial passages, and inhalation of vapors leads to destruction of tissue. Repeated contact of the skin, or especially the eyes, with the vapors, or the dilute acids (which may be harmless in single doses) may cause dermatoses (dilute phosphoric acid is not known to produce this effect). In general, *chronic* exposure to irritant vapors at low concentration produces as its most obvious clinical manifestations chronic exposure bronchitis and eye problems.

It should be mentioned that contact with vapors of strong acids or bases is often extremely hazardous. Nitric acid is a special case, as oxides of nitrogen (NO, NO_2, N_2O_3, and N_2O_4) are formed, reach the lungs, and, several hours later (after an asymptomatic period of perhaps 1 to as many as 48 hr, depending on the concentration breathed) produce some degree of respiratory distress, lassitude, cyanosis, headache, sometimes nausea, followed in severe cases by pulmonary edema and death. Fortunately, laboratory exposures rarely progress this far. In severe cases, if treated during the asymptomatic period, the pulmonary edema, and thus the fatal outcome, can be avoided. Nitric acid vapors must be treated with great respect.

d / Dehydrating Agents

A number of substances have sufficient affinity for water that they cause severe burns when in contact with the skin, irrespective of their other chemical properties. Concentrated sulfuric acid, sodium hydroxide, calcium oxide (CaO), and glacial acetic acid are among the dehydrating agents that are of importance in the laboratory. The charring of sugar by dehydration with concentrated sulfuric acid is a standard lecture demonstration.

e / Oxidizing Agents

These substances are not only corrosive to the skin, but are fire or explosion hazards as well should they come into contact with organic or inorganic reducing agents. (Remember, all hydrocarbons and most solvents are reducing agents.) Perchloric acid and chromic acid are two of the more important examples. Precautions for handling them will suggest methods for handling other similarly dangerous materials (some of these precautions repeat those given in the previous chapter for reactive chemicals).

i / Perchloric Acid

This substance ($HClO_4$) is available in purer form than other acids. At temperatures of 160°C it is also a strong dehydrating agent in concentrated or in vapor form. Anhydrous perchloric acid may be formed with stronger dehydrating agents such as phosphorus pentoxide or concentrated sulfuric acid. The anhydrous perchloric acid explodes immediately upon contact with any reducing agent.

The *Manufacturing Chemists Association Safety Manual* recommends that, in addition to the obvious precautions, all perchloric acid reactions be carried out in a hood behind safety shields. No more than 1-lb (454-g) bottles should be stored in a laboratory, and these in a padded container, with the bottle itself protected by a soft glass padding. Pouring from container to graduated cylinder should be carried out over a sink. After completion of a reaction, copious water should be used to wash the spent acid down the drain. The hood used should be metallic or ceramic, supplied with a strong exhaust to the roof, and accumulation of dust in or on the hood must not be permitted. The vapor must not be allowed into air conditioning systems.

When heating perchloric acid, use a sand bath, never an oil bath (a general precaution used for all strong oxidizing agents). Containers should be glass or teflon; corks or rubber stoppers are *absolutely* to be avoided. Spills are a particular problem: wood may absorb the acid, and later, if heated, may burn or explode. Spills must be mopped up with copious water, repeatedly, and the mop itself thoroughly rinsed.

ii / Chromic Acid

Chromium trioxide is added to sulfuric acid to produce chromic acid cleaning solution, which traditionally has been used to cut grease in volumetric ware. It is not often needed, because most glassware can be washed satisfactorily with a hot detergent solution. When chromic acid is dropped on the skin, it can cause long-lasting ulcers, not ordinary burns. Furthermore, chromium(VI) is a recognized carcinogen.

iii / Other Oxidizing Agents

Other substances, such as bromine, also attack eyes and skin. In general, acids, bases, dehydrating agents, and oxidizing agents are dangerous classes of chemicals, although not all members of a given class show identical behavior or require identical precautions.

5.4 CHRONIC EFFECTS AND PERMISSIBLE LIMITS

Your body can presumably survive a dose of one molecule of any-thing. On the other hand, even water can be somewhat toxic if one drinks several gallons each day. The question of dose of toxic sub-stances is obviously crucial to understanding the seriousness of a given exposure. Very toxic substances are those for which an expo-sure to "small" doses must be a matter of concern. Unfortunately, the criteria for determining dangerous doses are not always soundly based, nor are they necessarily useful in the laboratory situation. There are measures of acute and chronic toxicity. Many of the stan-dards were originally set to avoid acute exposure symptoms, with little regard for chronic effects. This is a situation that is now being remedied, although a completely adequate set of standards will not be available for many years. The standards that had been in effect up until 1971 were called **threshold limit values**. These had been issued for some 450 substances, obviously only a small fraction of those in use. The present standards are called **time weighted averages** (TWA), which is what the threshold limit values were also. The TWA is defined as "An employee's exposure . . . in any 8-hour work shift of a 40-hour work week, shall not exceed the 8-hour time weighted average for that material . . ." with the exposure to be computed according to

$$E = \frac{C_a T_a + C_b T_b + \ldots + C_n T_n}{8} \qquad (5-1)$$

where

E is the equivalent exposure for the working shift
C is the concentration during any period of time T where the concentration remains constant
T is the duration in hours of the exposure at the concentration C

The value of E shall not exceed the TWA. The values of the TWAs are given in Table 5.1, taken from Table Z–1 of OSHA 2077 (*General Industry Standards*). There are also ceiling values, and any substance preceded by the letter C "shall at no time exceed the ceiling value given for that material in the table." There are also some standards that are more complex, allowing brief excursions above a ceiling value. These are given in Table 5.2. Note that in Table 5.2, if exposure goes above the acceptable ceiling for a certain period, it must be averaged out by lower exposure sometime during the same work day. Another set of standards, applying to certain mineral dusts, is shown in Table 5.3. These standards have been in effect since January 1, 1976. However, standards are changing rapidly for some substances, such as benzene (benzene had a threshold limit value of 35 ppm in 1952. By the time the new law took effect in 1971, it was 10 ppm. As of May 21, 1977, it was to become 1 ppm; this limit has been legally challenged by the American Petroleum Institute. It has been made "permanent," however, effective March 1978; enforcement awaits court decision.) Furthermore, the standards given in the table are the legal limits; the National Institute of Occupational Safety and Health is now issuing "Criteria Documents" for many substances, recommending greatly reduced limits for an appreciable number of substances. These recommended limits are usually more realistic in terms of protecting health than the present TWAs and may, in many cases, become the legal limits after adoption by OSHA. Therefore, it is advisable to check with the most recent available documents before adopting procedures for the use of any substance.

Even the NIOSH standards are, for the most part, developed for healthy adults, and may or may not properly take account of the requirements of those who may be ill, may be particularly susceptible, or may be otherwise in a special situation (for example, pregnant women). Legal standards are often based on economic feasibility as much as on health considerations. (text continues on p. 81)

Table 5.1 ALLOWABLE EXPOSURE TO AIR CONTAMINANTS

Maximum exposure allowed by OSHA in an 8-hour work shift of a 40-hour work week
(time weighted averages). For substances preceded by C, exposure shall at no time exceed
the ceiling value given here.

Substance	p.p.m.[a]	mg./m^3[b]
Acetaldehyde	200	360
Acetic acid	10	25
Acetic anhydride.	5	20
Acetone.	1,000	2,400
Acetonitrile	40	70
Acetylene dichloride, see 1, 2-Dichloroethylene		
Acetylene tetrabromide.	1	14
Acrolein.	0.1	0.25
Acrylamide–Skin		0.3
Acrylonitrile–Skin.	20	45
Aldrin–Skin		0.25
Allyl alcohol–Skin	2	5
Allyl chloride	1	3
C Allyl glycidyl ether (AGE)	10	45
Allyl propyl disulfide	2	12
2-Aminoethanol, see Ethanolamine		
2-Aminopyridine	0.5	2
Ammonia.	50	35
Ammonium sulfamate (Ammate)		15
n-Amyl acetate.	100	525
sec-Amyl acetate.	125	650
Aniline–Skin.	5	19
Anisidine (o, p-isomers)–Skin		0.5
Antimony and compounds (as Sb)		0.5
ANTU (alpha naphthyl thiourea)		0.3
Arsenic and compounds (as As)		0.5
Arsine.	0.05	0.2
Azinphos-methyl–Skin		0.2
Barium (soluble compounds).		0.5
p-Benzoquinone, see Quinone		
Benzoyl peroxide		5
Benzyl chloride	1	5
Biphenyl, see Diphenyl		
Bisphenol A, see Diglycidyl ether		
Boron oxide		15
C Boron trifluoride	1	3
Bromine.	0.1	0.7
Bromoform–Skin	0.5	5
Butadiene (1, 3-butadiene)	1,000	2,200
Butanethiol, see Butyl mercaptan		
2-Butanone.	200	590

See footnotes at end of table.

Table 5.1 (continued)

Substance	p.p.m.[a]	mg./m³[b]
2-Butoxy ethanol (Butyl Cellosolve)–Skin	50	240
Butyl acetate (n-butyl acetate)	150	710
sec-Butyl acetate	200	950
tert-Butyl acetate	200	950
Butyl alcohol	100	300
sec-Butyl alcohol	150	450
tert-Butyl alcohol	100	300
C Butylamine–Skin	5	15
C tert-Butyl chromate (as CrO_3)–Skin		0.1
n-Butyl glycidyl ether (BGE)	50	270
Butyl mercaptan	10	35
p-tert-Butyltoluene	10	60
Calcium arsenate		1
Calcium oxide		5
Camphor	2	
Carbaryl (Sevin ®)		5
Carbon black		3.5
Carbon dioxide	5,000	9,000
Carbon monoxide	50	55
Chlordane–Skin		0.5
Chlorinated camphene–Skin		0.5
Chlorinated diphenyl oxide		0.5
Chlorine	1	3
Chlorine dioxide	0.1	0.3
C Chlorine trifluoride	0.1	0.4
C Chloroacetaldehyde	1	3
α-Chloroacetophenone (phenacylchloride)	0.05	0.3
Chlorobenzene (monochlorobenzene)	75	350
o-Chlorobenzylidene malononitrile (OCBM)	0.05	0.4
Chlorobromomethane	200	1,050
2-Chloro-1,3-butadiene, see Chloroprene		
Chlorodiphenyl (42 percent Chlorine)–Skin		1
Chlorodiphenyl (54 percent Chlorine)– Skin		0.5
1-Chloro, 2,3-epoxypropane, see Epichlorhydrin		
2-Chloroethanol, see Ethylene chlorohydrin		
Chlorethylene, see Vinyl chloride		
C Chloroform (trichloromethane)	50	240
1-Chloro-1-nitropropane	20	100
Chloropicrin	0.1	0.7
Chloroprene (2-chloro-1,3-butadiene)–Skin	25	90
Chromium, sol. chromic, chromous salts as Cr		0.5
Metal and insol. salts		1
Coal tar pitch volatiles (benzene soluble fraction) anthracene, benzo[a] pyrene, phenanthrene, acridine, chrysene, pyrene		0.2
Cobalt, metal fume and dust		0.1

See footnotes at end of table.

Table 5.1   (continued)

Substance	p.p.m.[a]	mg./m³[b]
Copper fume		0.1
Dusts and Mists		1
Cotton dust (raw)		1
Crag® herbicide		15
Cresol (all isomers)–Skin	5	22
Crotonaldehyde	2	6
Cumene–Skin	50	245
Cyanide (as CN)–Skin		5
Cyclohexane	300	1,050
Cyclohexanol	50	200
Cyclohexanone	50	200
Cyclohexene	300	1,015
Cyclopentadiene	75	200
2,4-D		10
DDT–Skin		1
DDVP, see Dichlorvos		
Decaborane–Skin	0.05	0.3
Demeton®–Skin		0.1
Diacetone alcohol (4-hydroxy-4-methyl-2-pentanone)	50	240
1,2-diaminoethane, see Ethylenediamine		
Diazomethane	0.2	0.4
Diborane	0.1	0.1
Dibutylphthalate		5
C o-Dichlorobenzene	50	300
p-Dichlorobenzene	75	450
Dichlorodifluoromethane	1,000	4,950
1,3-Dichloro-5,5-dimethyl hydantoin		0.2
1,1-Dichloroethane	100	400
1,2-Dichloroethylene	200	790
C Dichloroethyl ether–Skin	15	90
Dichloromethane, see Methylenechloride		
Dichloromonofluoromethane	1,000	4,200
C 1,1-Dichloro-1-nitroethane	10	60
1,2-Dichloropropane, see Propylenedichloride		
Dichlorotetrafluoroethane	1,000	7,000
Dichlorvos (DDVP)–Skin		1
Dieldrin–Skin		0.25
Diethylamine	25	75
Diethylamino ethanol–Skin	10	50
Diethylether, see Ethyl ether		
Difluorodibromomethane	100	860
C Diglycidyl ether (DGE)	0.5	2.8
Dihydroxybenzene, see Hydroquinone		
Diisobutyl ketone	50	290
Diisopropylamine–Skin	5	20

See footnotes at end of table.

Table 5.1 (continued)

Substance	p.p.m.[a]	mg./m³ [b]
Dimethoxymethane, see Methylal		
Dimethyl acetamide–Skin	10	35
Dimethylamine.	10	18
Dimethylaminobenzene, see Xylidene		
Dimethylaniline (N,N-dimethylaniline)–		
Skin.	5	25
Dimethylbenzene, see Xylene		
Dimethyl 1,2-dibromo-2,2-dichloroethyl		
phosphate, (Dibrom)		3
Dimethylformamide–Skin	10	30
2,6-Dimethylheptanone, see Diisobutyl ketone		
1,1-Dimethylhydrazine–Skin	0.5	1
Dimethylphthalate		5
Dimethylsulfate–Skin	1	5
Dinitrobenzene (all isomers)–Skin		1
Dinitro-o-cresol–Skin		0.2
Dinitrotoluene–Skin		1.5
Dioxane (Diethylene dioxide)–Skin. . . .	100	360
Diphenyl	0.2	1
Diphenylmethane diisocyanate (see		
Methylene bisphenyl isocyanate (MDI)		
Dipropylene glycol methyl ether–Skin . .	100	600
Di-sec-octyl phthalate (Di-2-ethylhexyl-		
phthalate).		5
Endrin–Skin		0.1
Epichlorhydrin–Skin	5	19
EPN–Skin		0.5
1,2-Epoxypropane, see Propyleneoxide		
2,3-Epoxy-1-propanol, see Glycidol		
Ethanethiol, see Ethylmercaptan		
Ethanolamine	3	6
2-Ethoxyethanol–Skin	200	740
2-Ethoxyethylacetate (Cellosolve acetate)–		
Skin.	100	540
Ethyl acetate.	400	1,400
Ethyl acrylate–Skin.	25	100
Ethyl alcohol (ethanol)	1,000	1,900
Ethylamine.	10	18
Ethyl sec-amyl ketone (5-methyl-3-		
heptanone).	25	130
Ethyl benzene	100	435
Ethyl bromide	200	890
Ethyl butyl ketone (3-Heptanone).	50	230
Ethyl chloride	1,000	2,600
Ethyl ether	400	1,200
Ethyl formate	100	300

See footnotes at end of table.

Table 5.1 (continued)

Substance	p.p.m.[a]	mg./m³ [b]
C Ethyl mercaptan	10	25
Ethyl silicate	100	850
Ethylene chlorohydrin–Skin	5	16
Ethylenediamine	10	25
Ethylene dibromide, see 1,2-Dibromoethane		
Ethylene dichloride, see 1,2-Dichloroethane		
C Ethylene glycol dinitrate and/or Nitroglycerin–Skin	[d]0.2	1
Ethylene glycol monomethyl ether acetate, see Methyl cellosolve acetate		
Ethylene imine–Skin	0.5	1
Ethylene oxide	50	90
Ethylidine chloride, see 1,1-Dichloroethane		
N-Ethylmorpholine–Skin	20	94
Ferbam		15
Ferrovanadium dust		1
Fluoride (as F)		2.5
Fluorine	0.1	0.2
Fluorotrichloromethane	1,000	5,600
Formic acid	5	9
Furfural–Skin	5	20
Furfuryl alcohol	50	200
Glycidol (2,3-Epoxy-1-propanol)	50	150
Glycol monoethyl ether, see 2- Ethoxyethanol		
Guthion ®, see Azinphosmethyl		
Hafnium		0.5
Heptachlor–Skin		0.5
Heptane (n-heptane)	500	2,000
Hexachloroethane–Skin	1	10
Hexachloronaphthalene–Skin		0.2
Hexane (n-hexane)	500	1,800
2-Hexanone	100	410
Hexone (Methyl isobutyl ketone)	100	410
sec-Hexyl acetate	50	300
Hydrazine–Skin	1	1.3
Hydrogen bromide	3	10
C Hydrogen chloride	5	7
Hydrogen cyanide–Skin	10	11
Hydrogen peroxide (90%)	1	1.4
Hydrogen selenide	0.05	0.2
Hydroquinone		2
C Iodine	0.1	1
Iron oxide fume		10
Isoamyl acetate	100	525
Isoamyl alcohol	100	360
Isobutyl acetate	150	700

See footnotes at end of table.

Table 5.1 (continued)

Substance	p.p.m.[a]	mg./m³ [b]
Isobutyl alcohol	100	300
Isophorone	25	140
Isopropyl acetate	250	950
Isopropyl alcohol	400	980
Isopropylamine	5	12
Isopropylether	500	2,100
Isopropyl glycidyl ether (IGE)	50	240
Ketene	0.5	0.9
Lead arsenate.		0.15
Lindane–Skin		0.5
Lithium hydride		0.025
L.P.G. (liquified petroleum gas)	1,000	1,800
Magnesium oxide fume		15
Malathion–Skin		15
Maleic anhydride.	0.25	1
C Manganese		5
Mesityl oxide	25	100
Methanethiol, see Methyl mercaptan		
Methoxychlor		15
2-Methoxyethanol, see Methyl cellosolve		
Methyl acetate	200	610
Methyl acetylene (propyne)	1,000	1,650
Methyl acetylene-propadiene mixture		
(MAPP)	1,000	1,800
Methyl acrylate–Skin	10	35
Methylal (dimethoxymethane)	1,000	3,100
Methyl alcohol (methanol)	200	260
Methylamine	10	12
Methyl amyl alcohol, see Methyl isobutyl carbinol		
Methyl n-amyl ketone (2-Heptanone). . .	100	465
C Methyl bromide–Skin	20	80
Methyl butyl ketone, see 2-Hexanone		
Methyl cellosolve–Skin	25	80
Methyl cellosolve acetate–Skin	25	120
Methyl chloroform	350	1,900
Methylcyclohexane	500	2,000
Methylcyclohexanol.	100	470
o-Methylcyclohexanone–Skin	100	460
Methyl ethyl ketone (MEK), see 2-Butanone		
Methyl formate.	100	250
Methyl iodide–Skin	5	28
Methyl isobutyl carbinol–Skin	25	100
Methyl isobutyl ketone, see Hexone		
Methyl isocyanate–Skin	0.02	0.05
C Methyl mercaptan.	10	20

See footnotes at end of table.

Table 5.1   (continued)

Substance	p.p.m.[a]	mg./m³[b]
Methyl methacrylate	100	410
Methyl propyl ketone, see 2-Pentanone		
C α Methyl styrene	100	480
C Methylene bisphenyl isocyanate (MDI)	0.02	0.2
Molybdenum:		
Soluble compounds		5
Insoluble compounds		15
Monomethyl aniline—Skin	2	9
C Monomethyl hydrazine—Skin	0.2	0.35
Morpholine—Skin	20	70
Naphtha (coaltar)	100	400
Naphthalene	10	50
Nickel carbonyl	0.001	0.007
Nickel, metal and soluble cmpds, as Ni		1
Nicotine—Skin		0.5
Nitric acid	2	5
Nitric oxide	25	30
p-Nitroaniline—Skin	1	6
Nitrobenzene—Skin	1	5
p-Nitrochlorobenzene—Skin		1
Nitroethane	100	310
Nitrogen dioxide	5	9
Nitrogen trifluoride	10	29
Nitroglycerin—Skin	0.2	2
Nitromethane	100	250
1-Nitropropane	25	90
2-Nitropropane	25	90
Nitrotoluene—Skin	5	30
Nitrotrichloromethane, see Chloropicrin		
Octachloronaphthalene—Skin		0.1
Octane	500	2,350
Oil mist, mineral		[e]5
Osmium tetroxide		0.002
Oxalic acid		1
Oxygen difluoride	0.05	0.1
Ozone	0.1	0.2
Paraquat—Skin		0.5
Parathion—Skin		0.1
Pentaborane	0.005	0.01
Pentachloronaphthalene—Skin		0.5
Pentachlorophenol—Skin		0.5
Pentane	1,000	2,950
2-Pentanone	200	700
Perchloromethyl mercaptan	0.1	0.8
Perchloryl fluoride	3	13.5
Petroleum distillates (naphtha)	500	2,000

See footnotes at end of table.

Table 5.1 (continued)

Substance	p.p.m.[a]	mg./m³ [b]
Phenol–Skin	5	19
p-Phenylene diamine–Skin		0.1
Phenyl ether (vapor)	1	7
Phenyl ether-biphenyl mixture (vapor)	1	7
Phenylethylene, see Styrene		
Phenyl glycidyl ether (PGE)	10	60
Phenylhydrazine–Skin	5	22
Phosdrin (Mevinphos ®)–Skin		0.1
Phosgene (carbonyl chloride)	0.1	0.4
Phosphine	0.3	0.4
Phosphoric acid		1
Phosphorus (yellow)		0.1
Phosphorus pentachloride		1
Phosphorus pentasulfide		1
Phosphorus trichloride	0.5	3
Phthalic anhydride	2	12
Picric acid–Skin		0.1
Pival ® (2-Pivalyl-1,3-indandione)		0.1
Platinum (Soluble Salts) as Pt		0.002
Propargyl alcohol–Skin	1	
Propane	1,000	1,800
n-Propyl acetate	200	840
Propyl alcohol	200	500
n-Propyl nitrate	25	110
Propylene dichloride	75	350
Propylene imine-Skin	2	5
Propylene oxide	100	240
Propyne, see Methylacetylene		
Pyrethrum		5
Pyridine	5	15
Quinone	0.1	0.4
RDX–Skin		1.5
Rhodium, Metal fume and dusts, as Rh		0.1
Soluble salts		0.001
Ronnel		10
Rotenone (commercial)		5
Selenium compounds (as Se)		0.2
Selenium hexafluoride	0.05	0.4
Silver, metal and soluble compounds		0.01
Sodium fluoroacetate (1080)–Skin		0.05
Sodium hydroxide		2
Stibine	0.1	0.5
Stoddard solvent	500	2,950
Strychnine		0.15
Sulfur dioxide	5	13
Sulfur hexafluoride	1,000	6,000

See footnotes at end of table.

Table 5.1   (continued)

Substance	p.p.m.[a]	mg./m³ [b]
Sulfuric acid		1
Sulfur monochloride	1	6
Sulfur pentafluoride	0.025	0.25
Sulfuryl fluoride	5	20
Systox, see Demeton ®		
2,4,5T		10
Tantalum		5
TEDP–Skin		0.2
Tellurium		0.1
Tellurium hexafluoride	0.02	0.2
TEPP–Skin		0.05
C Terphenyls	1	9
1,1,1,2-Tetrachloro-2,2-difluoroethane . .	500	4,170
1,1,2,2-Tetrachloro-1,2-difluoroethane . .	500	4,170
1,1,2,2-Tetrachloroethane–Skin	5	35
Tetrachloroethylene, see Perchloroethylene		
Tetrachloromethane, see Carbon tetrachloride		
Tetrachloronaphthalene–Skin		2
Tetraethyl lead (as Pb)–Skin		0.075
Tetrahydrofuran	200	590
Tetramethyl lead (as Pb)–Skin		0.07
Tetramethyl succinonitrile–Skin	0.5	3
Tetranitromethane	1	8
Tetryl (2,4,6-trinitrophenyl-methyl- nitramine)–Skin		1.5
Thallium (soluble compounds)–Skin as Tl		0.1
Thiram		5
Tin (inorganic cmpds, except oxides.		2
Tin (organic cmpds)		0.1
Titanium dioxide		15
C Toluene-2,4-diisocyanate	0.02	0.14
o-Toluidine–Skin	5	22
Toxaphene, see Chlorinated camphene		
Tributyl phosphate		5
1,1,1-Trichloroethane, see Methyl chloroform		
1,1,2-Trichloroethane–Skin	10	45
Trichloromethane, see Chloroform		
Trichloronaphthalene–Skin		5
1,2,3-Trichloropropane	50	300
1,1,2-Trichloro 1,2,2-trifluoroethane . . .	1,000	7,600
Triethylamine	25	100
Trifluoromonobromomethane	1,000	6,100
2,4,6-Trinitrophenol, see Picric acid		
2,4,6-Trinitrophenylmethylnitramine, see Tetryl		
Trinitrotoluene–Skin		1.5
Triorthocresyl phosphate		0.1

See footnotes at end of table.

Table 5.1   (continued)

Substance	$p.p.m.^a$	$mg./m^{3\,b}$
Triphenyl phosphate		3
Turpentine	100	560
Uranium (soluble compounds) . . . . :		0.05
Uranium (insoluble compounds)		0.25
C Vanadium:		
$\quad$V$_2$O$_5$ dust		0.5
$\quad$V$_2$O$_5$ fume		0.1
Vinyl benzene, see Styrene		
C Vinyl chloride	500	1,300
Vinylcyanide, see Acrylonitrile		
Vinyl toluene	100	480
Warfarin		0.1
Xylene (xylol)	100	435
Xylidine–Skin	5	25
Yttrium		1
Zinc chloride fume		1
Zinc oxide fume		5
Zirconium compounds (as Zr)		5

[a] Parts of vapor or gas per million parts of contaminated air by volume at 25°C. and 760 mm. Hg pressure.
[b] Approximate milligrams of particulate per cubic meter of air.
(No footnote "c" is used to avoid confusion with ceiling value notations.)
[d] An atmospheric concentration of not more than 0.02 p.p.m., or personal protection may be necessary to avoid headache.
[e] As sampled by method that does not collect vapor.

Table 5.2 ALLOWABLE EXPOSURE TO AIR CONTAMINANTS

For these substances OSHA allows a peak exposure above the acceptable ceiling for the time period shown (but average exposure for the entire day must not exceed the 8-hour time-weighted average).

Material	8-Hour Time Weighted Average	Acceptable Ceiling Concentration	Acceptable Maximum Peak Above the Acceptable Ceiling Concentration for an 8-Hour Shift.	
			Concentration	Maximum Duration
Benzene	1 p.p.m.	5 p.p.m.		
Beryllium and beryllium compounds	2 μg./m³	5 μg./m³	25 μg./m³	30 minutes.
Cadmium fume	0.1 mg./m³	3 mg./m³		
Cadmium dust	0.2 mg./m³	0.6 mg./m³		
Carbon disulfide	20 p.p.m.	30 p.p.m.	100 p.p.m.	Do.
Carbon tetrachloride	10 p.p.m.	25 p.p.m.	200 p.p.m.	5 minutes in any 4 hours.
Ethylene dibromide	20 p.p.m.	30 p.p.m.	50 p.p.m.	5 minutes.
Ethylene dichloride	50 p.p.m.	100 p.p.m.	200 p.p.m	5 minutes in any 3 hours.
Formaldehyde	3 p.p.m.	5 p.p.m.	10 p.p.m.	30 minutes.
Hydrogen fluoride	do			
Fluoride as dust	2.5 mg./m³			
Lead and its inorganic compounds	0.2 mg./m³			
Methyl chloride	100 p.p.m.	200 p.p.m.	300 p.p.m.	5 minutes in any 3 hours.
Methylene Chloride	500 p.p.m.	1,000 p.p.m.	2,000 p.p.m.	5 minutes in any 2 hours.

Table 5.2   (continued)

Material	8-Hour Time Weighted Average	Acceptable Ceiling Concentration	Acceptable Maximum Peak Above the Acceptable Ceiling Concentration for an 8-Hour Shift.	
			Concentration	Maximum Duration
Organo (alkyl) mercury	0.01 mg./m^3	0.04 mg./m^3		
Styrene	100 p.p.m.	200 p.p.m.	600 p.p.m.	5 minutes in any 3 hours.
Trichloro-ethylene	do	do	300 p.p.m.	5 minutes in any 2 hours.
Tetrachloro-ethylene	do	do	do	5 minutes in any 3 hours.
Toluene	200 p.p.m.	300 p.p.m.	500 p.p.m.	10 minutes.
Hydrogen sulfide		20 p.p.m.	50 p.p.m.	10 minutes once only if no other measurable exposure occurs.
Mercury		1 mg./10m^3		
Chromic acid and chromates		do^3		

Table 5.3　OSHA MAXIMUM ALLOWANCES FOR MINERAL DUSTS

Substance	Mppcf[a]	mg/m³
Silica:		
Crystalline:		
Quartz (respirable)	$\left(\dfrac{250^{b}}{\%SiO_2 + 5}\right)$	$\left(\dfrac{10mg/m^3}{\%SiO_2 + 2}\right)^{c}$
Quartz (total dust)		$\left(\dfrac{30mg/m^3}{\%SiO_2 + 2}\right)$
Cristobalite: Use ½ the value calculated from the count or mass formulae for quartz.		
Tridymite: Use ½ the value calculated from the formulae for quartz.		
Amorphous, including natural diatomaceous earth	20	$\left(\dfrac{80mg/m^3}{\%SiO_2}\right)$
Silicates (less than 1% crystalline silica):		
Mica	20	
Soapstone	20	
Talc (non-asbestos-form)	20[d]	
Talc (fibrous). Use asbestos limit		
Tremolite (see talc, fibrous)		
Portland cement	50	
Graphite (natural)	15	
Coal dust (respirable fraction less than 5% SiO_2		$\left(\begin{array}{c}10mg/m^3\\ \text{or}\\ \dfrac{2.4mg/m^3}{\%SiO_2 + 2}\end{array}\right)$
For more than 5% SiO_2		
Inert or Nuisance Dust:		
Respirable fraction	15	5mg/m³
Total dust	50	15mg/m³

Note: Conversion factors—
mppcf× 35.3 = million particles per cubic meter
　　　　　　 = particles per c.c.

[a]Millions of particles per cubic foot of air, based on impinger samples counted by light-field technics.

[b]The percentage of crystalline silica in the formula is the amount determined from airborne samples, except in those instances in which other methods have been shown to be applicable.

[c]Both concentration and percent quartz for the application of this limit are to be determined from the fraction passing a size-selector with the following characteristics:

[d]Containing < 1% quartz; if > 1% quartz, use quartz limit.

Even though we must take the standards with a grain of salt, the OSHA and NIOSH standards are the best available in the United States, and we list these. It should be understood that other countries have standards that are in many cases more stringent than United States standards. We should be prepared for the possibility that United States standards will be tightened as time goes on.

Another problem that is of considerable importance in the laboratory is that of mixtures of chemicals. In general, this is governed by the formula

$$E_m = \frac{C_1}{L_1} + \frac{C_2}{L_2} + \ldots \frac{C_n}{L_n} \tag{5-2}$$

where

E_m is the equivalent exposure for the mixture
C is the concentration of a particular contaminant
L is the exposure limit for that contaminant

This formula obviously assumes that there are no specific interactions among contaminants.

5.5 CLASSES OF CHRONICALLY TOXIC SUBSTANCES

At this point let us consider the chronic toxicity and carcinogenicity of certain compounds. [See also NIOSH publication #77–181, "Occupational Diseases: A Guide to Their Recognition," for a more detailed listing.]

a / Chlorinated Aliphatic Hydrocarbons

This designation covers a very wide range of compounds, including carbon tetrachloride and vinyl chloride. To some extent, however, the target organs and symptoms for many of these substances are similar. The degree of toxicity varies widely, however. The following paragraphs consider some specific compounds.

i / Carbon Tetrachloride (CCl$_4$)

This compound has relatively low *acute* toxicity, and was once used as a deworming agent, for which purpose a 3-mL dose was given orally. (In high enough dosage, it can cause nausea, diarrhea, kidney and liver damage, and finally death.) However, in 1976, a spill of carbon tetrachloride into the Ohio River, resulting in drinking water concentrations of a few hundred parts per billion, led the Environmental Protection Administration to impose emergency controls, including the shutting down of a large industrial plant.

This action was called for by our increased knowledge of the chronic effects of carbon tetrachloride, which is now known to be hepatotoxic in quite low doses, if repeated over time. Secondarily, kidney injury and visual disturbances can occur. Since carbon tetrachloride is a good organic solvent, it can remove skin oils, causing septic dermatitis. It is also absorbed through the skin. The effect on liver and kidneys is intensified by alcohol. At early stages, before the liver becomes swollen or malfunctions in a gross manner, it would be necessary to use liver function tests to detect the damage. Poisoning can occur by inhalation, ingestion, or skin absorption. Since the acute lethal dose for mice is approximately 1 per cent in air, only chronic poisoning is likely to be a serious problem. However, carbon tetrachloride, like chloroform (CHCl$_3$), does have some anesthetic properties at lower concentrations.

Precautions for handling carbon tetrachloride are much like those for other toxic organic compounds. Remember that the vapor, with a molecular weight of 152, is considerably denser than air, so that it will accumulate near the floor. This must be taken into account when estimating the efficacy of ventilation.

ii / Chloroform (CHCl$_3$)

This substance has been used as an anesthetic, and may be slightly less toxic than carbon tetrachloride. It has, however, been found to be carcinogenic, and in 1976 was banned by the Food and Drug Administration from use in drug, cosmetic, and food packaging applications. Concentrations above 20 ppm may produce toxic symptons, although the listed ceiling concentration (OSHA) is 50 ppm.

iii / 1,1,1-trichloroethane; methyl chloroform ($CH_3 CCl_3$)

This is a common solvent. It has narcotic effects, but is one of the least toxic of the chlorinated hydrocarbons.

iv / 1,1,2-trichloroethane ($CHCl_2 CH_2 Cl$)

This compound causes depression of the central nervous system, and is otherwise toxicologically similar to carbon tetrachloride, with about the same degree of toxicity; hence, it is much worse than 1,1,1-trichloroethane, which should be substituted for the 1,1,2 compound if possible.

v / Trichloroethylene ($Cl_2 C{=}CHCl$)

In addition to possible liver damage, the principal effects are on the central nervous system. Symptoms resemble ethanol intoxication, and chronic exposure can produce drug dependence (addiction). Withdrawal symptoms may include death due to ventricular fibrillation. Although liver damage has not been established at occupational levels, liver cancer has been shown in tests on mice.

vi / Vinyl Chloride ($CH_2 {=}CHCl$)

Until recently thought to be fairly innocuous, save for being narcotic in high concentrations, it is now known to be a carcinogen (causing angiosarcoma of the liver). Its TWA was dropped from 500 ppm to 1ppm. It also causes "vinyl chloride disease," which includes among its symptoms loss of feeling in the hands and feet, and slight anemia.

vii / Other Chlorinated Aliphatic Hydrocarbons

These compounds appear to vary considerably in their toxicity. All members appear to have some anesthetic properties and may be skin irritants owing to their solvent action. Central nervous system effects are variable in their severity, but at least some damage to kidney and especially liver is to be expected. After the vinyl chloride experience, it would be wise to treat all members of the group with caution, unless proven harmless.

Effects of all members of the group tend to be intensified by alcohol.

Other halogenated hydrocarbons are dealt with under aromatic hydrocarbons, and the respective elements. In general, toxicity increases as one goes down the periodic group; carbon tetrabromide (CBr_4) for example, is far more toxic than carbon tetrachloride.

b / Aromatic Hydrocarbons

These substances are also liver and kidney poisons, and they may in addition depress the respiratory system; in some cases they are central nervous system depressants and affect the blood-forming cells of the bone marrow as well. Some compounds require individual consideration.

i / Benzene

This compound is above all a danger to bone marrow and a cause of various blood disturbances, which in turn may lead to irreversible aplastic anemia as well as leukemia. For this reason the allowable TWA concentration of benzene has been abruptly lowered to 1 ppm, (pending resolution of legal challenge) which works out to about 1 drop in 10 m^3 volume of air. Also, and more relevant for student laboratories, there is a maximum allowable 5 ppm exposure for 15 min in the new standard. This can easily be exceeded in a 2- or 3-hr laboratory period. In addition to its effect on bone marrow, benzene shares with other aromatics, as well as many other hydrocarbons, a narcotic effect at high doses and the hepatotoxic effect discussed previously.

It is primarily because of the effect of benzene on the blood-forming tissue that it should never be used as a solvent for ordinary laboratory purposes (substitute toluene, cyclohexane, or, preferably, simple alkanes, which are far less toxic than either of the other possibilities; of course, if the toluene contains 5 per cent benzene impurity, it is not satisfactory, either). If benzene must be used, serious precautions are required. Like other organic solvents, it can pass through the intact skin. Obviously, it should be used only in a hood; an opened bottle should be stored only in a hood. If benzene is to be refluxed, it should not be allowed to vaporize too rapidly. Protective clothing may be required. Furthermore, persons under the age of 18 are reported to be particularly susceptible to bone marrow poisons, so benzene has no place in high school or most freshman laboratories. Pregnant women and nursing mothers are groups at particularly high risk from benzene.

Medical testing should be performed regularly on personnel with chronic exposure to benzene, and records must be kept for at least 20 years after exposure ceases (see Chapter 8).

There are many compounds that are as toxic as benzene, or worse. For lack of space, we shall not be able to discuss them in so much detail. However, it must be remembered that precautions just as stringent as these will be required. Furthermore, many compounds have not been as thoroughly investigated; it is very likely that many unpleasant surprises may be in store.

ii / Toluene and the Xylenes

Toluene does not appear to have the same insidious effects on the bone marrow as benzene, unless it contains benzene as an impurity. It does have an anesthetic effect and can cause central nervous system damage. The acute toxicity of toluene is a bit worse than that of benzene, though neither is an *acute* poisoning hazard of great severity (about 5 per cent toluene in air is fatal to 50 per cent of mice.) Prolonged exposure to 200 ppm of toluene causes fatigue, nausea, loss of appetite, vertigo and headache. Like other organic solvents, it must be detoxified in the liver and thus may cause alcohol intolerance.

Derivatives of toluene for the most part have similar toxicity. Xylene, however, may be worse; in addition to central nervous system damage, it may have a benzenelike effect on bone marrow. Chronic exposure may reduce resistance. Also, xylene may contain benzene as an impurity. All compounds in this class are organic solvents, and, like benzene, may enter via the intact skin. In addition, eczema due to defatting skin is a possible problem.

iii / Naphthalene ($C_{10}H_8$)

A severe eye and skin irritant (primarily eye), naphthalene appears otherwise much less toxic than benzene. Also, naphthalene has a much lower vapor pressure. (It is a solid whose melting point, 80°C, is about equal to the boiling point of benzene.) In general, although not entirely safe, it can be used more freely than the single ring compounds discussed previously. The dust must be suppressed; also, melted naphthalene must be used only in a hood to avoid dangerous vapor concentrations.

iv / Anthracene ($C_{14}H_{10}$)
This substance should not be spilled on the skin. Although the pure compound is not as toxic as might be expected, it does have a serious effect on the skin and is to some extent a photosensitizer. That is, skin contaminated with anthracene would be damaged by sunlight, or ultraviolet light. However, the impurities typically found in anthracene are far more deadly [see paragraph (v)].

v / Larger Aromatic Ring Systems
These substances are found in coal tar, chimney soot, and cigarette smoke and are the oldest known occupational carcinogens. Some of the most powerful carcinogens known are included in this group (3,4-benzanthracene and methylcholanthrene, among others). Most of these are also more powerful photosensitizers than anthracene. It is important not only to avoid skin contact but to prevent release of dust containing these compounds into the laboratory atmosphere. These substances obviously are to be avoided in undergraduate laboratories and should be treated as are other carcinogenic substances. They are present not only as impurities in anthracene, but in the tars resulting from other reaction mixtures.

c / Chlorinated Aromatic Hydrocarbons
This category includes some of the most widely distributed commercial compounds: DDT, the herbicides 2,4-dichlorophenoxyacetic acid (2,4-D), 2,4,5-trichlorophenoxyacetic acid (2,4,5-T), and the polychlorinated biphenyls (PCBs). In the United States, DDT and the PCBs have been ruled out of commercial use. The use of the herbicides has been restricted as their toxicity and environmental effects have become more clearly understood. Similarly to nonaromatic chlorinated hydrocarbons, they are toxic to liver and kidneys and somewhat neurotoxic in many cases; some also cause a peculiar skin condition called chloracne.

Impurities in the synthesis of these compounds may be a particular problem. Dioxin, possibly the most toxic synthetic molecule known, may be formed accidentally in the synthesis of trichlorophenol (as in the disaster at Seveso, Italy, in 1976) and of 2,4-D and 2,4,5-T. The problem is especially serious if the reaction overheats. Therefore, extreme caution is required in preparing student experiments involving syntheses in which unwanted chlorinated aromatics may appear. In the case of dioxin, even nanogram quantities may be extremely serious.

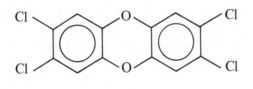

dioxin

d / Aliphatic Amines

The lower aliphatic amines are often found in normal body tissues. Furthermore, their odor gives clear warning of their presence. Most are irritants, particularly to the eyes, and some may damage the respiratory tract. Solutions of some tend to be sufficiently basic to be irritants for that reason alone. Other than this, these compounds for the most part are more easily metabolized or excreted unchanged than most other classes of organic chemicals. However, see hydrazine, azo, and azoxy compounds in paragraphs (i) and (viii) of part (e) and part (i).

e / Aromatic Amines

These compounds tend to be used as synthetic intermediates, and the impurities coming from the syntheses are often more dangerous than the major product.

i / Aniline

This substance is a serious acute poison, producing cyanosis due to methemoglobinemia. It can produce the accompanying symptoms of vertigo, headache, and mental disorientation. Chronic exposure leads to anemia and loss of appetite and weight, as well as skin lesions. The pure compound apparently is not carcinogenic, but its typical impurities and reaction products cause bladder cancer.

Aniline is fat soluble, and readily absorbed through the skin, as are other single ring derivatives. From this point of view, the hydrochloride, which is water soluble and therefore not absorbed through the skin, is safer to handle. However, the hydrochloride is a powder and, if dispersed as a dust, may be inhaled, producing the same effects as the free base. It is reported that a 1 per cent solution of methylene blue (10 cm^3, intravenous) is a nearly specific antidote for acute poisoning by aniline. Oxygen administration may also help. In case of poisoning by direct contact, the victim should undergo complete and *immediate* washing with soap and tepid water (fingernails included). The victim's clothing is likely to be contaminated and should be removed and laundered.

Other aromatic amines, including some found as impurities in aniline, are carcinogenic, especially to the bladder. For many years, aniline was itself thought to be responsible. However, the substances responsible are impurities, particularly naphthylamines and aminodiphenyls. Diazo dyes are also dangerous, and may be formed from aniline or other aromatic amines.

ii / Naphthylamines

These compounds are known and powerful bladder carcinogens; at least, 2-naphthylamine is. There is some question as to whether the carcinogenicity of 1-naphthylamine is due to 2-naphthylamine as an impurity. Use of naphthylamines has been discontinued industrially because of their carcinogenicity.

iii / 4-Aminodiphenyl and 4,4'-Diaminodiphenyl (Benzidine)

Both substances are too dangerous for industrial use; they are potent carcinogens and can be absorbed through the skin. Both compounds are solid at room temperature but have appreciable vapor pressures. The carcinogenicity of some of their derivatives is open to question, but all should be treated with great caution until proven safe.

iv / p-Phenylenediamine

This is a skin sensitizer and respiratory sensitizer, and may cause asthma in sensitized individuals.

v / Chloroanilines

These substances are somewhat similar to aniline; they are absorbed through the skin, are potent methemoglobin formers, are irritating to the eyes, and so on.

vi / Diphenylamine

This compound is not known to be extremely toxic, nor is it considered to be carcinogenic; however, carcinogenic impurities, especially 4-aminodiphenyl, may be present when it is synthesized and may be concentrated in the tar fraction when diphenylamine is distilled.

vii / Nitroanilines

These compounds are extremely powerful agents of methemoglobin formation and are easily absorbed through the skin. They may cause hemolysis, and chronic exposure causes liver damage. Contact with skin, eyes, or clothing must be avoided, and in general the same personal decontamination procedure as for aniline is to be followed. As with aniline, get medical attention in the event of eye contact.

The salts, especially the hydrochloride, being water soluble, offer the same tradeoff in hazards as aniline does; in place of being absorbed through the skin, the salts may be dispersed as powders.

viii / Hydrazine (N_2H_4) and Substituted Hydrazines

These substances are extremely strong reducing agents, and therefore present a danger of fire or explosion in air (see Chapter 4). They are also volatile and can be absorbed through the skin. Toxic effects include methemoglobinemia, hemolysis, anemia, bone marrow damage, liver, kidney, and heart damage, central nervous system symptoms, depression, and various acute symptoms, the exact effects depending on the particular substitutional derivative. Hydrazine itself and dimethyl hydrazine are reported to be carcinogenic. All hydrazines should be handled with good local ventilation, or in a closed system. Protective clothing with vinyl or neoprene coating should be used. Personal decontamination is similar to that for aniline. In case of eye contact, *immediately* begin flushing the eyes and continue for at least 15 min. TWAs for hydrazines are 1 ppm or below.

f / Aliphatic Hydrocarbons

These are fire and explosion hazards (see Chapter 4). Methane and ethane are pharmacologically inert. (They would be asphyxiants by excluding oxygen in the concentrations needed to produce toxic effects.) Higher saturated hydrocarbons are intoxicants and central nervous system depressants in concentrations of thousands of parts per million. However, liquid hydrocarbons are solvents and do dissolve skin oils, causing dermatitis. If liquid hydrocarbons are aspirated into the lung, pneumonitis leading to a fatal outcome may result. n-Hexane has recently been reported to cause peripheral neuritis.

g / Aliphatic Alcohols, Ketones, Aldehydes, Ethers, and Organic Acids

Again, there is a fire and explosion hazard associated with these compounds (see Chapter 4). The category covers such a wide range of compounds that generalizations are obviously of limited value. Those members that are solvents dissolve skin oils and cause dermatitis. Many have central nervous system effects, especially as anesthetics, depressants, or intoxicants. Some ketones (for example, methylbutyl ketone) appear to cause severe peripheral neuritis. The most common members, other than their effect on skin, are not as toxic as many of the members of the categories discussed previously. Before working with the less common members of these groups, it would be well to check available toxicity data.

h / Amides

The amide functional group is not itself particularly toxic. This is hardly surprising, since it is the fundamental linkage in proteins. Amides of fatty acids are less dangerous than many other compounds discussed here. However, a number of common amides are not so harmless, and some, including acetamide and thioacetamide, as well as dimethylformamide (DMF) and dimethylacetamide (DMAC), are reported to be hepatotoxic. The latter two compounds cause various gastrointestinal effects and may be absorbed through the skin. Acrylamide monomer causes both serious peripheral and central nervous system damage, with the primary route of entry through the skin. Those amides with alkylating nitrogens may be carcinogenic.

i / Nitro and Azo Compounds
Some are carcinogenic, including 4-nitrobiphenyl and certain diazo compounds, including the diethyl and dinaphthyl compounds. See also part e (vii), page 89.

j / N-Nitroso Compounds
As a general rule, these should be regarded as carcinogenic, unless proven otherwise.

k / Nitrogen Mustards
These compounds are alkylating agents, some of which are sufficiently toxic to have been used as war gases. Some mustards have been used as chemotherapeutic agents for cancer, and some cause cancer. The toxic properties of any nitrogen mustard should be checked before being used in the laboratory.

l / Epoxy Compounds
These substances may produce dizziness, nausea, and unconciousness from acute exposure to fumes. Epoxy glues are used for many purposes and have the same hazardous properties. Prolonged exposure, especially skin contact, may produce sensitization, preventing *all* future work in the vicinity of epoxies. In case of skin contact, use disposable wipes to remove syrupy liquids, then remove contaminated clothing under the safety shower. **Do not use solvents to remove epoxy.** Eye contact requires flushing the entire eye surface for 15 min. In each of the above cases, medical attention is required.

Low temperature curing materials are generally more hazardous than high temperature materials. If room temperature epoxies are heated while curing, vaporized solvents and catalysts are even more hazardous. Wastes from epoxy glues, if not fully cured, should be disposed of with the hazardous wastes.

m / Methyl Fluorosulfonate (Magic Methyl®)

This methylating agent is one of the most dangerous compounds commercially available. The distributor points out that animal tests indicate an LD_{50} of 5 to 6 ppm in air (this is the dosage lethal to 50 per cent of test animals, and is not to be confused with a TLV or TWA). The substance also causes death upon skin contact. Precautions include rubber gloves, (but be sure the rubber is not of a type which will be penetrated by the compound!) eye protection, and ventilation. Most important is proper training of personnel who will use the compound.

n / Natural Products

A number of natural products are carcinogenic, or otherwise toxic. Among these are aflatoxin, found in peanuts, and cycasine, safrole, isosafrole, and pyrrolizidine alkaloids.* Most antibiotics, some of which can cause hemolysis and other serious physiological effects, must be handled with extreme care. Space limitations prevent listing the degree of hazard of each of the antibiotics that might be of interest to some chemists, but toxic properties should be checked before working with any of these substances or any other substance known to be physiologically active.

o / Polymers and Plastics

Polymers, per se, are generally innocuous; once polymerized, substances are sufficiently unreactive as to preclude toxic effects. However, plastics and polymers may, in principle at least, be hazards in three ways.

i / Monomers

Many monomers are toxic. Vinyl chloride, the monomer of polyvinyl chloride (PVC), causes angiosarcoma of the liver, as discussed

---

*List according to Rademacher (*J. Chem. Educ.* **53**:757, 1976.)

in section 5.5 (a) (vi), p. 83. Workers preparing their own polymer may become exposed. There may also be some monomer remaining in the PVC, depending on the method of preparation. Other significantly toxic monomers include isocyanates, of which toluene diisocyanate (TDI) and methylene bisphenyldiisocyanate (MDI) are probably the most important. These (especially TDI) are reacted with polyhydroxy compounds to form polyurethanes. They have ceiling (15 min average) allowable concentrations of only 0.02 ppm, which suggests that they are toxic in concentrations much lower than those accepted for most organic compounds. Sensitization and loss of lung function are the most serious problems.

Other common monomers, such as styrene or ethylene, are less toxic.

ii / Plasticizers
Appreciable quantities of some plasticizers are vaporized from some plastics, if they are allowed to become too warm. This is occasionally a problem in a confined space.

iii / Decomposition Products
Depending on the polymer and the conditions of decomposition, these may be extremely dangerous both as acute and as chronic toxins. Decomposition products are not limited to the monomers, but include oxidation products, and, if decomposition occurs with sufficient heat, a variety of nitrogen compounds as well (assuming decomposition in air). Recalling that soot is carcinogenic, and that the decomposition of polyurethane may produce cyanides, caution in the use of polymers and plastics under conditions in which they may burn or decompose becomes necessary. Similar considerations apply to the tars that form in some organic reactions.

p / OSHA and NIOSH Lists
A brief consideration of two lists of carcinogens is in order at this point. One is a group of 17 substances listed by OSHA, the other a list of 1500 substances considered to be probable carcinogens by NIOSH. The OSHA list of "Cancer Suspect Agents" is as follows: 4,4'-methylenebis(2-chloroaniline) (MOCA); benzidine and its salts; 1-naphthylamine; 2-naphthylamine; 3,3'-dichlorobenzidine and its

salts; 4-aminobiphenyl (all of the preceding are primary amines); 4-nitrobiphenyl; ethylenimine (both of which can be readily converted to primary amines); N,N-dimethylaminoazobenzene; bis (chloromethyl)ether; chloromethyl methyl ether; N-nitrosodimethylamine; 2-acetylaminofluorene; 2-propriolactone; vinyl chloride; asbestos dust; and benzene. The last three have been added since the initial list was issued. The NIOSH list of 1500 suspected carcinogens is available as *Suspected Carcinogens: Subfile of the Toxic Substances List.* The differences between the short OSHA list and the much longer NIOSH list seem to have to do with the fact that OSHA issues regulations that must withstand court challenge from industries which would lose money if the substance were ruled carcinogenic. (OSHA is now preparing a "generic" classification scheme; it will be challenged legally.) It would be surprising if great numbers of the NIOSH 1500 were not also human carcinogens; simple prudence would dictate treating them as carcinogens until proven otherwise.

q / Heavy Metals and Their Compounds
Most heavy metals are well known toxins. Some have carcinogenic properties, may attack the liver somewhat and the kidneys seriously, and a number are neurotoxic.

i / Mercury
This metal deserves pride of place because of its ubiquity coupled to its utility. When these factors are combined with the difficulty of handling mercury, it becomes the leading problem in this category. It has been recognized as an occupational hazard for about a century; the Mad Hatter in *Alice in Wonderland* displayed typical mercury poisoning symptoms (hatters used mercuric nitrate in sizing felt).

Mercury is readily absorbed via the respiratory tract, and it easily spills and flows (hence the old name, quicksilver). Chronic effects include severe kidney and neurotoxicity (including tremors and personality changes), as well as swelling of gums and excessive saliva-

*GPO Stock # 1733–00084, $4.85, from Superintendent of Documents, GPO, Washington, D.C. 20402.

tion. Alkyl mercury and phenyl mercury compounds may be absorbed through the skin, and the alkyl mercury compounds may cause permanent brain damage. Both sets of compounds cause severe delayed skin burns. Inorganic salts of mercury may be extremely corrosive to the mucous membranes (For example, mercuric chloride, $HgCl_2$).

The vapor pressure of metallic mercury is 2×10^{-3} torr at $25°C$. This corresponds to about 2.5 ppm, or more than 20 mg/m^3. The acceptable occupational ceiling level is 0.1 mg/m^3 (0.05 mg/m^3 in California). To contaminate a 300-m^3 laboratory to a level of 0.1 mg/m^3 requires the evaporation of just 30 mg of Hg, or about 2.2 $\times 10^{-3}$ cm^3 of the liquid metal. Obviously, 1 cm^3 of Hg, if spilled, represents a problem.

Spill prevention and cleanup involve some simple precautions. In pouring mercury, in addition to always using funnels, it is well to place a basin or large beaker underneath the vessel or tube into which the mercury is being poured. This does not guarantee spill immunity, but it may at the least reduce the clean-up problem greatly. Also, the pouring should be done over a stainless steel or plastic surface, with rims but no crevices. To prevent volatilization, the surface should not be excessively warm. Clean-up problems are often most serious because of the loss of small droplets in crevices. Commercial mercury clean-up kits, which include pads for picking up droplets of mercury, are available. Droplets may also be picked up by suction into a trap (a suction flask will do), through a glass tube whose open end has been drawn down to about 2 mm. If any mercury is suspected of remaining behind, the area may be dusted with flowers of sulfur, which may then be collected.

A method of cleaning up mercury by freezing it with dry ice and acetone has also been reported. The frozen mercury can then be picked up.* Old analytical laboratories in particular should be inspected for mercury in crevices; frequently, droplets remain behind for years.

Testing for mercury may be carried out by atomic absorption; unlike other metals, there is no need for heating, as the vapor pressure is sufficient. Other specific mercury monitors are also available.

As previously mentioned, organic mercury compounds may be absorbed through the skin and cause irritation. Phenyl mercury

---

*J. Chem. Educ., **50**:739 (1973).

compounds are at least as toxic as inorganic mercury. Alkyl mercury compounds are much worse. The chronic neurological effects are incurable. Acute poisoning may be treated by injection of British Anti-Lewisite (BAL), but this is used only in case of imminent danger to life. Permissible TWAs are generally 0.01 mg/m^3, one tenth the metallic mercury level.

ii / Chromium
Chromium(VI) salts and chromic acid cleaning solution produce long-lasting ulcers on the skin that reach the bone eventually. Dust may cause ulceration of the nasal septum. Healing takes many months. Long exposure to chromium (state uncertain) in dust leads to lung cancer. In case of exposure to chromium(VI), skin should be carefully washed, avoiding friction and, if possible, sweating. Skin cuts, *no matter how slight,* must be washed immediately and treated with a 10 per cent CaNa$_2$ EDTA ointment. This reduces chromium(VI) to chromium(III), which the remaining EDTA can chelate.

iii / Cadmium
This substance is about as toxic as mercury, although the symptoms are different. The TWA of 0.1 mg/m^3 very likely should be lowered. Cadmium fumes (including those of cadmium oxide (CdO), cadmium sulfide (CdS), and other inorganic compounds) are the primary hazard. Kidney damage and emphysema are the most serious symptoms of chronic exposure. No specific treatment exists. Cadmium is also a carcinogen. Traces may be present in soldering fumes.

iv / Arsenic
Arsenic has been used as a poison for so many centuries that it has become famous for its toxic properties. A number of organic As compounds were developed as chemical warfare agents; one was Lewisite,

$$
\begin{array}{cc}
\text{H} & \text{H} \\
| & | \\
\end{array}
$$
$$
\text{Cl-C=C-AsCl}_2 ,
$$

for which BAL was developed as an antidote. Arsine (AsH$_3$), an extraordinarily toxic gas, is produced by contact of arsenic acid

($HAsO_3$) with zinc, so that accidental generation is a serious possibility. For example, arsenic acid may contact the zinc lining of a galvanized bucket, releasing hydrogen gas, which reacts with arsenic acid to produce arsine. Reactions of arsenic with water must also be considered. Hydrolysis of metallic arsenides produces arsine. Finally, chronic exposure to arsenic compounds is now considered to cause cancer. As a result, the TWA for arsenic and its compounds is now 0.002 mg/m^3 ; 1 mg contaminates a 500 m^3 laboratory.

v / Nickel

It has been estimated that 5 per cent of *all* eczema comes from contact with nickel or nickel compounds (including watches, coins, pins, and so on). Fortunately, nickel does not appear to be too toxic when ingested. However, inhalation of nickel dust and the dust of nickel compounds may cause pneumonitis with adrenal cortical insufficiency and lung cancer.

One compound, nickel carbonyl [$Ni(CO)_4$], deserves special attention. It is toxic at extremely low levels, with a TWA of 0.001 ppm. The primary effects are on the lungs. The poisoning is insidious, and symptoms may appear days after exposure, although 12 to 36 hr is more common. These delayed symptoms may be extremely severe; beginning with constrictive chest pain, severe pulmonary and sometimes gastrointestinal symptoms ensue. Death may follow in 4 to 11 days. Nickel carbonyl may be inadvertently formed if carbon monoxide (CO) contacts an active form of nickel (as during welding of stainless steel). The compound is volatile and can be used only in an enclosed system; obviously it has no place in ordinary laboratory procedures. There is an antidote (diethyl dithiocarbamate, Dithiocarb); victims should be removed to medical attention immediately.

Both nickel and nickel carbonyl are believed to be carcinogenic. Rats exposed to fairly large single doses of nickel carbonyl developed lung cancer about 2 years later; the same effect was found from repeated small doses.

With other nickel compounds, the most obvious precaution is to avoid skin contact. Wear gloves when working with salts of nickel, and, should contact occur, wash thoroughly. Anyone who has developed nickel allergy, as shown by dermatitis, should be removed from contact with the metal or its compounds.

vi / Cobalt

This is more likely to be an industrial than a laboratory hazard. Cobalt compounds cause fibroses and granulomas of the lung; however, the principal route of entry is inhalation of dust, rather than contact with solutions of cobalt compounds. Cobalt has also been reported to be an allergen, and may show cross sensitization with nickel.

vii / Manganese

Manganese(II) compounds cause chronic poisoning. Higher oxidation states are caustic. The chronic poisoning causes nervous system and possibly pulmonary symptoms. Again, the primary problem is inhalation of dust.

viii / Lead

This is one of the best known toxic metals. Again, inhalation of dust is the principal route of entry, although ingestion is also a possibility. Fumes of the pure metal may be a problem if lead is heated above 500°C; ordinary soldering does not reach this temperature, but welding, silver soldering, and other processes do. Symptoms of lead poisoning include gastrointestinal disturbances, followed by serious neurological manifestations. Because the gastrointestinal symptoms appear first, poisoning may be misdiagnosed as colic or appendicitis. Anyone exposed to lead fumes must be given regular medical tests for blood lead levels.

ix / Vanadium

Vanadium trioxide (V_2O_3) and especially vanadium pentoxide (V_2O_5) are the hazards of primary importance. These may be inhaled; they are also toxic on skin contact. Single exposures may produce serious results, including bronchitis and pneumonia. A greenish discoloration of the tongue also results. Again, this is primarily a dust problem.

x / Selenium and Tellurium

The oxides of these metals are problems, as are the hydrides hydrogen selenide (H_2Se) and hydrogen telluride (H_2Te). Liver and kidney damage are the most important consequences of exposure to the oxides. The preliminary warning symptom is a strong garlic stench on the breath. Hydrogen selenide and hydrogen telluride smell worse than hydrogen sulfide. Anyone who has inhaled selenium fumes, hydrogen selenide, selenium oxide (SeO_2), or corresponding tellurium compounds, should be given oxygen therapy for 1 hr prophylactically, followed by hospitalization. Exposed skin should be washed with 5 per cent sodium thiosulfate ($Na_2S_2O_3$) solution, followed by use of 10 per cent sodium thiosulfate cream. Eye exposure should be treated by washing with 10% sodium thiosulfate solution, or with plain water.

xi / Beryllium
Beryllium enters almost entirely by inhalation and causes a chronic and often fatal lung disease, sometimes mistaken for tuberculosis. The disease often has cardiac and renal complications and causes joint pain. It is treated, although not necessarily successfully, with large doses of steroids.

As the disease results from the inhalation of any of a large number of beryllium compounds, all work with any beryllium compound should be considered hazardous, and all necessary precautions should be taken to suppress dust. Protective clothing is important as the dust can stick to clothes; the protective clothing must not be removed from the laboratory. There have been "neighborhood" cases, in which those in the vicinity (1 km) of beryllium workers, their clothes, and anything else that might carry dust, have suffered beryllium disease. Those who contract the disease in this manner show a mortality rate of 50 per cent, compared to 26 per cent for worker cases. Medical records of those exposed must be kept for 40 years. The permissible TWA is 0.002 mg/m^3.

xii / Bismuth
Although the toxic effects are qualitatively similar to those of lead, arsenic, or mercury, much higher doses are required to produce these effects. Precautions required are therefore much simpler, and normal housekeeping procedures should be sufficient to deal with clean-up.

xiii / Thallium
Thallium and its compounds are chronic poisons of about the potency of cadmium or mercury. Poisoning produces neurological symptoms, followed by massive loss of hair. Precautions are similar

to those for other very toxic dusts, including the use of suitable protective clothing. TWA is 0.1 mg/m^3.

Note that for the heavy metals (excluding mercury) and their compounds, the principal danger is exposure to dust. Since most of their compounds are handled in dry form, the risk is not trivial. If it is possible to weigh or otherwise handle heavy metal compounds in fairly large pellets, the risk can be minimized. Fine powders should be handled with gloves, in the hood, and care must be taken to avoid dispersing them into the laboratory atmosphere. For the most toxic compounds a totally closed system, similar to that recommended for handling radioactive substances, may be required (see Chapter 6). Suitable protective clothing must be used, and stringent rules for personal hygiene observed.

r / Halogens

i / Fluorine (F$_2$)

Fluorine, oxygen difluoride, and chlorine trifluoride are *extremely* strong oxidizing agents. At low concentrations they are irritating; at slightly higher concentrations, they have extremely corrosive effects on human skin. Protective clothing and rigorous personal hygiene help to reduce hazard and lanolin may be used as a barrier cream. However, the primary line of defense must be the proper enclosure of the system in which the compounds are used. If they are used in a vacuum line, in which there exists possibility of leakage, local exhaust should provide ventilation and respirators should be available (either self-contained breathing apparatus or pressure-demand types).

Hydrofluoric acid (HF) causes severe burns in dilute solution or vapor. It is crucial to insure adequate ventilation, to insure that safe levels (1 ppm) are not exceeded. *All* contact of vapor or liquid with eyes, skin, respiratory system, or digestive system must be avoided. Protective equipment, fabricated from inert plastics such as neoprene or polyvinyl chloride, should completely cover arms and legs, hands, feet, face, and body. Safety showers and eye wash fountains should be nearby. Finally, anyone working with hydrofluoric acid should be well trained in its hazards and in proper protective measures.

Fluorides are of only limited toxicity by ingestion. However, fluoroacetic acid and its salts are extremely toxic; they should be removed from the stomach if swallowed, and monoacetin (100 cm^3 in 500 cm^3 water) administered orally. Hospital treatment is required.

ii / Chlorine (Cl$_2$)

The pure element was used as a war gas in World War I. It has chronic effects on pulmonary function even at concentrations below its TWA of 1 ppm. Chlorine should be handled only in closed systems. As with fluorine, self-contained breathing apparatus or pressure-demand respirators should be available in case of a leak. Exposure to 40 to 60 ppm may cause serious injury, and 100 ppm may be lethal. Even a few inhalations may be fatal at 1000 ppm. Chronic exposure to 5 ppm creates a predisposition to tuberculosis, and may lead to chloracne, just as chlorinated aromatic hydrocarbons do.

Oxides of chlorine should be handled in the same manner as chlorine. Sulfur chloride and sulfur dichloride are intensely irritating fuming liquids that cause skin burns. They should also be used only in closed systems with protective clothing and respirators available. The same is true for thionyl chloride, sulphuryl chloride, and chlorsulphonic acid.

iii / Bromine (Br$_2$)

Molecular bromine, a liquid at room temperature, causes severe, long-lasting skin burns that may ulcerate. The vapor can cause acute or chronic poisoning, even 30 mg/m^3 producing dangerous acute symptoms. Chronic effects include neurological symptoms, cardiovascular disease, and thyroid dysfunction, among others. Obviously, bromine should be used only in closed systems, or with strong and reliable local ventilation. Precautions mentioned for chlorine should be observed. It should be noted that bromide salts are less stable than chlorides, and may decompose, releasing the element.

Organobromine compounds are generally far more toxic than the corresponding organochlorine compounds. Bromoform requires more care in handling than chloroform, and so on. Aromatic bromine compounds are particularly dangerous to the central nervous system. Polybrominated biphenyls (PBBs) were recently added to cattle feed by error in Michigan. People who drank the milk produced by the cattle lost coordination, memory, and showed other significant central nervous system symptoms; the corresponding PCB exposure would not have been as bad. Even bromobenzene produces neurological symptoms (among others).

iv / Iodine (I_2)

The element has a ceiling concentration of 0.1 ppm (1 mg/m^3), indicating a serious risk at lower concentration than bromine. Even 0.1 ppm may be irritating to the eyes. Iodine vapor (I_2) is irritating to the respiratory tract, and may cause pulmonary edema. Skin contact produces burns, which may ulcerate. The most serious chronic effect is on those suffering from thyroid disorders. Chronic exposure also leads to "iodism", a disease with a variety of symptoms.

Closed systems should be used when handling iodine, or at least strong local ventilation. Gloves should be worn to prevent skin contact.

Iodine compounds are generally more toxic than the corresponding bromine compounds, possibly because of the effect on the thyroid. It is just as well that iodine is sufficiently expensive that the industrial use of its compounds is discouraged.

s / Phosphorus and Its Compounds

Elemental phosphorus is a fire and explosion hazard as well as a toxic problem. White phosphorus burns spontaneously in air, and, if it comes in contact with the skin, causes extremely painful burns that may require that the affected area be cut away. On chronic ingestion or inhalation white phosphorus causes "phossy jaw," a chronic bone necrosis particularly affecting the jaw. Red phosphorus, if mixed with an oxidizing agent such as $KClO_3$, may explode. Furthermore, a number of phosphorus compounds can also cause very serious toxic problems.

i / Phosphine (PH_3)

Unlike ammonia, very small doses of phosphine can be lethal (TWA = 0.3 ppm). The compound has a very foul fishy odor, which may provide warning in time. However, the *lethal* dose is only 60 ppm for rats. The substance may be prepared by the action of acid or water on metallic phosphides, so working with metallic phosphides requires precautions to insure that water does not contact the material.

ii / Organophosphorus Compounds

These substances, which include some commonly used insecticides, exhibit various degrees of antiacetylcholinesterase action. In fact, the nerve gases synthesized at the time of World War II were in this class, the most potent being Sarin,

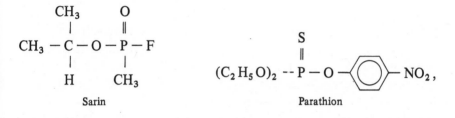

Sarin   Parathion

for which the lethal dose may be as low as 0.01 mg/kg. A chemical relative used as an insecticide, Parathion, is "merely" highly toxic. Symptoms include anorexia, nausea, vomiting, diarrhea, excessive salivation, pupillary constriction (often the first symptom to be noticed, as the room appears to go dark), bronchoconstriction, muscle twitching, convulsions, coma, and death by respiratory failure. Most serious for those occupationally exposed, effects are cumulative, and anyone working with acetylcholinesterase inhibitors should be tested for acetylcholinesterase activity. Absorption through the skin can be a serious problem (TWA (skin) = 0.11 mg/m^3), and sprays must also be avoided. First aid in case of acute poisoning may require *hours* of artificial respiration. Antidotes [atropine and PAM (pyridine-2-aldoxime methiodide)] and medical attention should be available if work is to be done with these or other acetylcholinesterase inhibitors.

Many other compounds have similar activity, though the level of toxicity may differ (Malathion, for example, is probably about 100 times less toxic than parathion.) In general, those working with phosphate esters should be aware of the danger of acetylcholinesterase inhibition, since major precautions are required.

iii / Inorganic Phosphorus Compounds

Many inorganic phosphorus compounds are quite corrosive, or strong irritants, or both—phosphorus pentoxide, phosphorus pentafluoride, and phosphorus pentachloride (TWA = 0.5 ppm), among others. In addition to phosphine, some compounds that are extremely toxic include phosphorus trioxide and phosphorus tribromide (which is also corrosive). Some compounds decompose on contact with oxygen or water giving toxic products. For example, phosphorus pentasulfide decomposes on contact with water to give hydrogen sulfide and on contact with air to give phosphorus pentoxide.

t / Vapors Associated with Atomic Absorption Spectroscopy

Atomic absorption spectroscopy (AA) is an analytical method for determining trace concentrations of various elements. The technique involves vaporizing the sample, so that the element would be dispersed in the laboratory if ventilation were not provided. When a sample contains any of the elements described in the preceding sections, whether or not it is the one for which the analysis is being carried out, ventilation is needed.

5.6 MONITORING FOR UNSAFE VAPOR CONCENTRATIONS

It does little good to know that a certain concentration of the vapor of a substance is unsafe, unless the actual concentration is in fact determined. For many purposes, it may be sufficient to work in the hood, confident that occasional exposure to a substance will not cause chronic effects. However, this is not sufficient for technicians, for faculty, for graduate students, or for industrial chemists, all of whom repeatedly come into contact with the same substances. Since some very common substances (carbon tetrachloride, benzene) may be very toxic, this is not a trivial consideration. Furthermore, even student exposures, although not covered by OSHA, may exceed OSHA limits, and presumably be unsafe. For example, according to the proposed emergency benzene standard, there is a 5 ppm ceiling for 15-min exposure. Other substances also have ceiling limits, beyond which even short term exposure is considered hazardous, although insufficient to produce acute symptoms. Monitoring can be done so as to produce instantaneous samples (grab samples), which give the concentration at one time and place. Sometimes average samples are needed, in which case sampling is carried out at one location over a fixed time interval and the total contaminant collected is then analyzed. Continuous monitoring is also possible.

a / "Grab" Samples

One convenient, though not very accurate, instrument for vapors is the "universal tester." This instrument, about the shape of a bicycle pump, although smaller in size, pulls a sample of air through a glass tube that contains an indicator. The sample of air is of calibrated volume (assuming the pump is used correctly), and the glass tube will show an indicator color change. The tube may be 10 to 15 cm long; the length of indicator that changes color is a measure of concentration. Different tubes are available for different substances, over 100 in all. For some substances, several tubes are available, to cover two or more ranges of concentration (for example, for carbon monoxide, carbon dioxide, ammonia, and hydrogen sulfide). The reproducibility of most tubes is not adequate for quantitative testing, but it is good enough to decide whether the concentration is safely under the limit, close to the limit, or significantly above the limit. Considering that the standards are not themselves exact, (except legally) this may be adequate for deciding whether control measures are justified or not.

The Mine Safety Appliances Universal Tester Pump kit without tubes costs $139. (as of 1978). The cost of tubes varies with the substance being tested, but is roughly $1 per test in most cases. Other companies such as Bendix make comparable equipment.

Other methods of taking grab samples also exist. Various types of containers (cans, suitably lined bags, round bottom flasks), may be evacuated, and opened at the location to be tested. The container may then be resealed, the contents concentrated if necessary (for example, by adding a solvent) and analyzed by gas chromatography or other appropriate technique. The procedure obviously has some difficulties, and requires considerable work to ensure proper calibration. Such problems as the possibility of adsorption of sample on the walls of the vessel or loss of sample by leakage suggest that such techniques can only be used with great care.

b / Sampling for a Finite Period

A sampling train with midget impingers or gas-washing bottles can be used to obtain samples over a finite time at a fixed location. In this system, air is pumped for a known length of time through a train, as illustrated in Figure 5.1. The amount of the pollutant of interest is

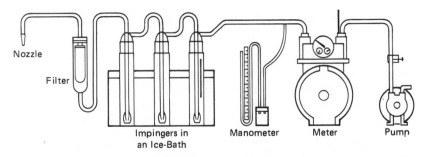

Figure 5.1   Particulate sampling train.

then determined by standard means. The sampling train must include a gas flow meter; flow rate multiplied by sampling time gives the total volume of air sampled, so that the amount of contaminant captured in the train can be referred back to an air concentration. This is the basis for many standard methods used in air pollution control measurements. Given the apparatus of a university chemistry department, or an industrial laboratory, it may be a useful method of monitoring laboratory pollutants as well. The result is an average over the time measured at the location of the intake nozzle.

c / Continuous Samples

Assuming an appropriate location is determined for sampling, continuous samples may nevertheless be required. This will be the case particularly for any substance for which a ceiling concentration, rather than a TWA, is given. For that matter, instantaneous samples cannot establish whether the TWA has been exceeded, unless they are taken so frequently that they approximate a continuous sampling. Continuous monitors typically may be fixed-wavelength infrared instruments with built-in alarms that are activated if a preset concentration is exceeded. Electrochemical instruments and paper-tape samplers can also be used as monitoring instruments. Gases for which continuous monitors are commercially available include sulfur dioxide (SO_2), nitrogen oxides (NO_x), hydrogen cyanide (HCN), tetraethyl lead, mercury, ammonia, fluorine, nickel carbonyl [$Ni(CO)_4$], chlorine, and hydrazine (N_2H_4). Manufacturer's catalogs should be consulted for specifications of particular instruments.

Wilks Scientific Co. manufactures a portable infrared spectrometer designed for pollutant monitoring. Using it, concentrations may be determined at any time or location. A portable gas chromatograph for monitoring is available from Analytical Instrument Development, Inc. Mine Safety Appliances, Bendix, and Matheson are among the firms manufacturing continuous monitoring instruments.

d / Personal Monitors

These devices, which perform much the same function as radiation dosimeters, are also available commercially from, among others, Anatole J. Sipin Co. and Du Pont. Personal monitors consist of a small constant-flow rate pump, usually worn on the belt or in the shirt pocket, and a charcoal tube, or other type sorbent tube, placed near the wearer's breathing zone (for example, clipped to the collar) and connected via plastic tubing to the pump. At the end of the sampling period, the charcoal tube is connected to the appropriate analytical instrument, such as a gas chromatograph, and then heated to release the trapped toxins, which can then be determined. Calibration with a standard sample would be required. This type of sampler would meet OSHA requirements for tests of personnel exposure. Perhaps 200 substances can be monitored using this type of sampler. The cost is several hundred dollars for the instrument, plus about $1 per test.

e / Validity of Samples

Each of the methods mentioned so far suffers from the limitation of being taken at a given spot, although the personal monitor does give the actual personal exposure. Except in the last two cases, exposure is not averaged over time but taken instantaneously. Although the personal monitor can be used to obtain an average, it will not reveal possibly dangerous surges. (Some forms of continuous monitors do respond to surges.)

This leads to the problem of obtaining suitably representative samples. Some vapors are more dense than air and tend to settle towards the floor. If one wishes to determine whether the air is approaching the lower explosive limit (see Chapter 4), it makes sense to sample near the floor. On the other hand, a toxic vapor, with inhalation as principal route of entry, would more reasonably be monitored in the breathing zone at about the height of mouth or nose. The distribution of personnel in the room, laboratory, or other workspace is obviously important. It does little good to know that the corridor is clean if the personnel are in a room. Because of irregularities in air currents, the appropriate location at which to obtain a representative sample is not always obvious on simple observation and may require a careful inspection or series of tests. It is well to remember that air flow may vary with the season, as the ventilation changes, as hoods are turned on or off, and so on.

5.7 AVOIDING EXPOSURE

The general methods for avoiding exposure have been discussed several times in general terms. Here we discuss some particulars.

a / Substitution

In some cases, less toxic materials may be used in place of more toxic—toluene in place of benzene, 1,1,1-trichloretheylene in place of 1,1,2-trichlorethylene, and so on. Care is required to insure that the material substituted is, in fact, less toxic than the original material. It is sometimes possible to substitute for a material on the NIOSH list of 1500 suspected carcinogens a substance on which few toxicological data are available; later information may show this to be a case of leaving the frying pan for the fire. In making the substitution, it is necessary to choose a material about which sufficient data are available to allow a rational choice of precautions in handling.

Nevertheless, a well-considered substitution may lower risk greatly. In particular, since the choice of less common substances for student laboratories is at least somewhat arbitrary, care should be taken to avoid unnecessary exposures to carcinogens or other seriously toxic materials, as it is virtually impossible to institute adequate precautions in such a laboratory. Students can be counted on to produce spills, to clean up inadequately, to touch chemicals with bare hands, and so on.

Another method of substitution is to avoid reaction paths that use carcinogenic intermediates.* In general, when planning safety precautions for a reaction, not only the intended final products and intermediates but possible side reactions and impurities should be considered. The example of dioxin in the synthesis of 2,4-D and 2,4,5-T, mentioned earlier, is a case in point.

Assuming that a seriously toxic material is to be used, it is well to consider whether it will be used in large concentrations or quantities, and used repeatedly, or used in a single experiment, or in micro quantity.

b / Enclosure

The use of a glove box to enclose a reaction is discussed in some detail in Section 6.3, part (g), page 128, in discussions of containment of radioactive materials. The same principles and techniques apply to confinement of nonradioactive toxic materials.

c / Ventilation

i / Hoods

Although the use of a glove box and enclosed system is required for the most serious toxins, most operations can be carried out in a properly functioning hood. In chemical reactions, the fumes or vapors are typically released at essentially zero velocity, and laboratory air is fairly still. This is the easiest case to ventilate. One measures the efficacy of a hood by determining the flow rate of air at the face of the hood. For most purposes 50 to 100 ft/min (15 to 30 m/min) is sufficient; however, if there are particularly toxic substances, a higher flow rate may be required. Also, secondary air currents may be a problem; the velocity of the flow at the hood must be adequate to overcome any drafts or other air turbulence that may bring material back to the laboratory. In addition to the linear flow rate, sufficient contaminated air must be removed so that adequate make-up air is brought in for breathing, if the operator is to work

*This problem has been discussed by Rademacher (*J. Chem. Educ.* **53**:757 (1976).

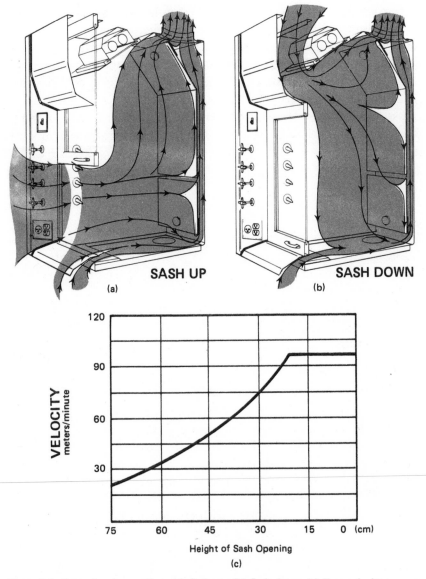

SASH UP

(a)

SASH DOWN

(b)

120

VELOCITY
meters/minute

90

60

30

75   60   45   30   15   0 (cm)

Height of Sash Opening

(c)

Figure 5.2 Fume hood operation. (a) Sash up. (b) Sash down. (c) Face velocity vs. open height. Data refer to "Airflow Supreme" design of Kewaunee Scientific Equipment Corp., Adrian, Michigan.

with his or her head in the hood. With proper hood design, this is possible.

The linear flow rate at the face of the hood is determined in part by the height of the sash. When the flow rate is measured, the sash should be at working height or above; if it is lower, a higher flow rate will be measured than is available when the hood is in use. A marking showing the maximum sash height at which the hood can be safely used should be placed at the side of the hood. An alarm may even be attached if the hood sash is raised too high. The flow rate should be measured at least twice a year, since hoods may clog, become dusty, or otherwise be reduced in efficiency. Also, adding hoods to an existing system can be a serious problem. The fans are generally not designed for the added load and will not provide adequate air velocity.

Hoods, of course, must be designed so that the air flow does not go through the operator's breathing zone (see Figure 5.2). This is an obvious point, and modern commercial laboratory hoods are usually designed correctly in this respect. Some ventilation designs do overlook it, though.

ii / General Ventilation

This is rarely a problem with which the chemist can be concerned; it does, however, arise in the design of new laboratories. The question is beyond the scope of this book, and a suitable work on industrial hygiene should be consulted. Similar considerations, sometimes with a more stringent standard, arise in the design of clean rooms. Considerations involve volume and direction of air flow; direction, source, and volume of make-up air; turbulence and secondary air currents; effects of windows, and so on.

General ventilation is of particular importance in the case of laboratories that are to be designed or redesigned for use with radioisotopes or carcinogens on a regular basis and in larger than microcurie or milligram quantities. In this case, the entire laboratory may have to be under negative pressure, and various other specific design features incorporated into the facility.*

*See J. N. Keith, "How to Design a Building Safe Against Hazards," *Int. J. Occup. Health Safety*, 46(2):46–58 (Mar/Apr 1977).

5.8 RESPIRATORS

In the chemistry laboratory, a respirator is strictly emergency apparatus. The choice of respirator does not appear to be an easy one, because of the impossibility of predicting which substance(s) it will have to protect against. The Standards Completion Program Committee, comprised of industrial hygienists from OSHA and NIOSH, has prepared a Respirator Decision Logic (RDL) to make possible the determination of an appropriate respirator for any given substance. Unfortunately, if the substance is not known in advance, the RDL cannot be applied. It is then necessary to have available a respirator that can handle a variety of extremely toxic vapors and fumes, some of which may not even be stored in the laboratory (for example, phosgene may be produced from carbon tetrachloride during a fire).

Under the circumstances, personnel entering an area with unknown contaminants almost certainly will require either self-contained breathing apparatus, with positive pressure in the face piece, or a pressure-demand respirator with auxiliary self-contained air supply, also giving positive pressure. The "gas mask" type of respirator with activated charcoal is not considered reliable, as the chance of a leak in the seal of the mask to the face is too great (also, anyone with a beard cannot use this type, or any type with a face piece that is supposed to seal, and requires a hood or helmet type respirator). The respirator with self-contained air supply provides positive pressure in the face mask so that any leaks go from inside out.

In case of emergency, it is necessary to have someone available who is able to use the equipment. Therefore, it is imperative that some of the laboratory staff be trained to use the apparatus and that sufficient respirators be on hand in each location in the building. Furthermore, cylinders of compressed air last, at most, 30 min; therefore, adequate numbers must be available. All cylinders must be checked regularly to insure that the contents have not been lost by leakage. Some masks may be used with building compressed air supplies; however, these may be rendered useless in an emergency by cutoff of the compressor or contamination of the supply. Also, the air must be known to be of breathable quality. Finally, the tubing, which must be carried behind, is even more of a hazard than the tanks of a self-contained unit. In fact, masks attached to building compressed air cannot be used in an area from which it may be impossible to escape without the apparatus. Hence, the requirement for self-contained respirators.

Very toxic substances, such as benzene at higher concentrations, require one of the following types of respirators: Type C supplied-air respirators operated in pressure-demand mode or other positive-pressure mode; or respirators with a full facepiece, hood, or helmet operated in a continuous-flow mode; or, alternatively, self-contained breathing apparatus, with full facepiece, operated in positive-pressure mode. In addition to these types, there are also specific masks (for example, canisters that specifically protect against chlorine). However, as it is not possible to predict which particular hazard, or combination of hazards, will be faced in an emergency, and the concentrations generated may be too high for the specific mask, it is necessary to have respirators of at least one of the types mentioned and to train personnel to use those types.

Only in unusual cases is a laboratory or department justified in keeping any other type of respirator on hand. Unless there is a specific need for a particular type, the cost of annual replacement alone would justify reliance on self-contained or supplied-air types, which are of general utility. Thus, the choice of respirators becomes considerably simpler.*

5.9 PROTECTIVE CLOTHING OTHER THAN GLASSES

Unlike respirators, which, for the chemist, are strictly emergency equipment, there are some forms of protective clothing that are regular necessities in some laboratories.

a / Gloves

Since skin contact is the primary route of entry for many toxic substances, gloves may often supply important protection. Rubberized gloves may be penetrated by some solvents. However, disposable gloves that offer protection are available from laboratory supply

---

*See the booklet *NIOSH Certified Equipment,* available from Publications, DTS, NIOSH, 4676 Columbia Parkway, Cincinnati, Ohio, 45226, for a list of manufacturers of approved respirators.

companies. Plastic gloves (for example, of polyvinyl chloride) cost 8 to 12¢ per pair. Reusable laboratory gloves (gauntlet length) offer protection against acids, alkalis, some solvents, and powders. Other reusable gloves are made specifically for use with solvents. Most have grips so that they can be used for reasonably delicate operations. It is necessary to choose gloves that offer protection against the particular hazard that will be encountered and still allow flexibility for laboratory manipulations. Gloves may not be needed routinely in undergraduate laboratories but are an important safeguard in many research situations. Even in undergraduate laboratories, if such substances as benzene, carbon tetrachloride, chloroform, aniline, nitrobenzene, nitroaniline, or similarly toxic agents are in use, gloves are an important safety precaution.

b / Lab Coats
To the extent that contact with certain solvents is a hazard, lab coats offer some protection. However, the traditional cloth coat is a protection primarily for clothing; it may even create a hazard for the wearer. Aniline, for example, if spilled and absorbed by the cloth, may be a continuing hazard until the coat is carefully washed. Since most lab coats are not washed regularly, the health risk may be greater than if no coat were worn. However, plastic coats, which can cover or replace the ordinary lab coat and be rinsed off or discarded, are often preferable. Although solvents may eventually penetrate these coats, they do offer at least some protection.

c / Shoe Coverings and Flooring
Disposable plastic shoe coverings afford some protection against spills. However, when solvents spill, the nonconductive plastic may increase the risk of spark. Where flammable vapors are present, conductive soles reduce dangers associated with static electricity. Laboratory surfaces, or flooring, that reduce danger from spills is available but conductive flooring does not appear to be commercially available.

d / Caps
Disposable plastic caps to protect hair may also be useful and are available.

5.10 CLEANING UP SPILLS

One case in which any or all of the protective gear described in the preceding section may be required is the clean-up of spills. Several varieties of hazard may be involved: the spill may be an acid or base, a strong oxidizing or reducing agent, aqueous or not, or a dehydrating agent, a very toxic solvent such as benzene, or a relatively nontoxic but flammable solvent such as octane. Mercury was considered in Section 5.5, part (p) (i), and flammable spills in Chapter 4.

a / Acids and Bases

Unnecessary personnel should be evacuated. Those who remain to clean up should put on plastic shoe coverings and plastic gloves. If the spill is on a bench top, a plastic coat would also be advisable.

Generally, the principal danger is from the corrosive properties of the material itself. However, with oxidizing acids, there is a danger of combustion if the spill should contact organic or other reducing substances. With nitric acid there is also a probable toxicity hazard, as oxides of nitrogen may be formed. In this case, a respirator may be required.

The size of the spill also partially determines the extent of the required precautions. A few milliliters of hydrochloric acid can be wiped away with a sponge, without any special equipment. If there is a large acid spill, however the area should be roped off and if possible, combustible materials and active metals should be removed (to prevent hydrogen generation). If an appropriate neutralizing material is available, such as soda ash, or sodium carbonate (Na_2CO_3) for acids or sodium bisulfate ($NaHSO_4$) for bases, it should be used. If none is available, it may be necessary to dilute the acid with water. This has the important disadvantage of spreading the spill. However, dilute acid can generally be mopped up, whereas concentrated acid would be too dangerous. If acid or base spills may be anticipated, it is possible to have available an appropriate neutralizer plus an absorbent (such as vermiculite), which together would make it possible to collect the spill with broom and dustpan in a form in which it can be safely disposed of without spreading. Acid and base spill kits are now available from several companies with all necessary components in prepared form.

b / Toxic Materials

i / Materials That Are Not Corrosive
If the chemical spilled is toxic but presents no reactive or corrosive problem, a procedure appropriate to the particular material may be used. A procedure developed for primary aromatic amines at Los Alamos will do for an example.* The steps are:

- Evacuate the area
- Have clean-up personnel don protective clothing, including respirator.
- Vacuum the contaminated area, using a specially constructed cleaner that filters exhaust air through charcoal and other amine filters. An ordinary vacuum cleaner may be modified for this purpose. Vacuuming removes the bulk of the spill without spreading it, as mopping would.
- Scrub the area with a solution that will remove the residue left after vacuuming (for primary amines, use methanolic 6 M hydrochloric acid or concentrated HCl diluted 1:1 with an aqueous solution of detergent).
- Test for presence of residue. For aromatic primary amines, two spot tests are needed. Ehrlich's reagent gives a yellow to orange color with 200 ng/cm^2 (nanograms per square centimeter) of aromatic amine. Fluorescamine is a more sensitive spot reagent, giving a fluorescent response. However, this may be lost due to fluorescence quenching. Hence, both tests must be negative to be sure the spill has been cleaned up.

Obviously, if a procedure like this is to be carried out, the equipment must be available, including protective gear, specially modified vacuum cleaner, and spot reagents. If one is intending to work with severely toxic or carcinogenic materials, safety planning must be an integral part of the experimental plan.

It is also easier to clean up a spill if the surface onto which the spill occurs is smooth and impervious. Floors, in particular, should be free of crevices that can trap spilled chemicals.

*R. W. Weeks, Jr., B. J. Dean, and S. K. Yasuda, *Int. J. Occup. Health Safety,* 46 (2):19 (Mar/Apr 1977).

ii / Toxic and Corrosive Materials

There is little else one can do than proceed as for the case of a toxic material. However, if the spill is corrosive and of appreciable size, it may be necessary to neutralize or dilute it before using a vacuum cleaner. Alternatively, a vacuum cleaner that is constructed of glass and teflon up to a cold trap large enough to remove the pollutant may be required. What is most important is to plan before beginning; this includes having designed and prepared equipment for clean-up in case of accident.

5.11 STORAGE OF TOXIC CHEMICALS

Storage of chemicals may be a problem not only from the point of view of reactivity (for example, ethers), but also because of toxicity. Just as chemicals may be incompatible from the point of view of reactivity, they may be incompatible because of their toxicity. An example of an incompatible pair would be potassium cyanide (KCN) and any acid, especially any strong acid, because this combination produces the extremely toxic gas, hydrogen cyanide.

A first step toward proper storage is classification according to risk. Alphabetical storage is no more satisfactory here than with reactive or flammable chemicals. Those chemicals that create little hazard (cholesterol; common salt—NaCl) may be stored with little ventilation, whereas those that present major dangers require great care. The same is true of incompatible pairs of chemicals, where separation may be an adequate solution. Vapor pressure is another important consideration; mercury may not be more toxic than arsenic, but its vapor pressure makes storage of the element much more of a problem than storage of elemental arsenic. Fine powders may not present the same problem as high vapor pressure, but transferring powders may produce dangerous quantities of dust.

For substances that are deemed to be sufficiently hazardous (recognized or presumptive carcinogens, mercury, and other substances of like degree of hazard), storage facilities should be chosen according to considerations of vapor pressure, mutual compatibility, ease of disposal, techniques required for handling, and quantity. Many of the same precautions required to ensure that high concentrations of the vapor of flammable liquids do not build up are appropriate for toxic materials as well. The toxic substances may have lower volatility; nevertheless, very much lower concentrations of the toxic materials can be tolerated. A ventilated storage cabinet should be used. However, the required ventilation rate need only be sufficient to keep the concentrations more or less within ceiling limits (no one spends 8 hr a day with his or her head in the cabinet).

The required ventilation rate depends also on leakage rate and volatility. It has been found* that once a reagent bottle has been opened, sufficient material will remain on the threads of a screw cap to allow finite quantities to evaporate, even if the cap reseals the bottle itself. If the cabinet is well ventilated, it remains effectively under negative pressure so that the room in which it stands should not be significantly contaminated. Of course, when a bottle or sample is taken from the cabinet, there must be proper transfer facilities or the room will become contaminated. A combination of monitoring and spot tests may be used to ascertain that levels are safe. To design ventilation systems, an estimate of the air flow required to prevent significant contamination of the room in which the storage cabinet stands must be made; at this time there does not appear to be a simple rule for doing so. Significant contamination would occur, according to the OSHA rule, if the *sum* of the ratios of concentrations to allowable TWAs exceeds unity (see Eq. 5-2, page 81).

The ventilation system itself must be filtered before the air is released outside. The problem is essentially the same as that encountered in disposing of toxic wastes. For a discussion, see Chapter 7. Alternatively, consider sealed storage with activated charcoal, as is provided by the "Static Hood," sold by Bel-Art Products or, for example, an old refrigerator containing activated charcoal. Stockroom personnel should wear disposable protective clothing.

If small bottles must be filled from large bottles, the procedure obviously requires protective clothing, especially gloves, and must be carried out in a hood. It is useful as well to have suitable pipettes or other means to carry out the transfer. Pipettes with disposable tips are useful and safe for quantities small enough to be transferred in this manner. Larger quantities require normal glassware, which must be cleaned carefully afterward. If the transfer is quantitative, the

*A Turk, E. Solomon, and H. Mark, *Amer. Lab.* Jan. 1970, p. 59.

waste disposal problem created by washing is limited. The person doing the washing must be suitably protected, of course.

Storage of gases in cylinders follows rules that are basically similar, regardless of whether or not the gases are toxic. The cylinders must be strapped securely to desk or wall, capped if not in use, and remote from any source of heat (radiators, ovens, burners, and so on). The gauge, if attached, should read zero pressure when the gas is not in use. (See Chapter 3 for details.) However, as soon as the gas is to be used, there is a considerable difference between toxic and nontoxic gases. Obviously, such gases as nitrogen, oxygen or argon may be safely vented into a room without toxic effect. It is equally obvious that this is not true for chlorine or nitric oxide (NO) or similarly dangerous gases. The system into which such a gas is to be transferred must be known in advance to be leakproof. Sufficient local ventilation must be available in case of any accident, so that the results will not spread outside the laboratory in which the leak occurs. Self-contained or positive-pressure-demand respirators should be available for the personnel in the affected area, should these be necessary.

Finally, at the end of such an experiment, it is necessary to remove the gas from the experimental system; this must also be planned in advance, with a train of gas-washing bottles to absorb the gas prepared and the gas still in the system. See Section 6.3 on the handling of radioactive gases and Chapter 7 on cleaning up.

6 Physical Hazards

6.1 CLASSIFICATION OF PHYSICAL HAZARDS

Physical hazards include radiation, both ionizing and nonionizing, as well as noise, heat, and cold. This list is more or less in order of importance for chemists. Noise may be the most serious industrial health problem in the United States, but it is a relatively minor problem for chemists. Heat and cold are rarely serious problems for chemists, either. However, ionizing radiation can be of considerable importance, with nonionizing radiation not very far behind.

Nonionizing radiation includes ultraviolet (uv), visible, infrared (ir), and microwave radiation. Ionizing radiation of importance to chemists principally includes x rays and the α, β, and γ rays from isotopes that may be handled in the laboratory.

6.2 NONIONIZING RADIATION

a / Ultraviolet

Excessive exposure to ultraviolet produces the symptoms of sunburn; uv may also seriously damage the eyes, and this is by far its most serious hazard in the laboratory. The exposed eye feels normal for 6 hr or more, then feels as though it had sand in it. The eye becomes irritated, and the condition painful. Chronic uv effects include loss of vision.

Laboratory uv sources should be enclosed. Most often, one has no reason to look directly at uv light, which cannot in any case be easily or directly detected by the eye. Instead, one looks at the fluorescence produced by the uv, as in detecting spots in thin-layer chromatography. Since uv, except at wavelengths above 330 nm, is stopped by plate glass but the visible fluorescence is not, plate glass makes an adequate safety barrier for uv light for most purposes. Spectroscopic sources (D_2 lamps, for instance) are generally enclosed, the light being directed into a monochromator. Such sources as high pressure mercury lamps, which may be used in photochemistry research labs, must be treated with care to avoid subjecting those working in the laboratory to uv from scattered or reflected light. (The explosion hazard connected with these lamps also requires that the lamp be enclosed). Glasses should provide adequate protection while the light source is aligned, but goggles may be needed if the source is intense and light may enter the eye from the side. Precautions for uv lasers are similar to those for visible or infrared (ir) lasers. Ultraviolet produces ozone (O_3). If the odor of ozone is detected adequate ventilation must be provided. Oxides of nitrogen may also be a problem.*

b / Visible and Infrared—The Laser

For the laboratory worker, a serious problem is presented by the laser. The intense beam of light can be focussed by the lens of the eye onto the retina, burning it, and thus causing permanent blindness. Even small continuous wave lasers, such as the 632.8 nm helium-neon laser available even to hobbyists, can produce this effect. Reflections from mirrors or pieces of metal are essentially as dangerous as the beam itself.

Aligning the beam produces the most serious risk. When seeking to align the beam, one should use dull, diffuse reflectors. Care must be taken to avoid turning the beam on unintentionally. It may be appropriate to leave the plug out until it is time to turn the beam on. Also, neither the laser nor any mirrors in the optical path should be left loose enough to swing around accidently, which might result in reflecting the beam onto others in the lab. Laser safety goggles,

*See NIOSH criteria document No. 1733–00012 on ultraviolet radiation.

which absorb light strongly at particular wavelengths but are effectively transparent elsewhere, also exist. However, they are not efficient enough to permit the beam to shine directly into one's eye, even with helium-neon lasers. They do greatly reduce the intensity of light and afford some protection, but the protection is limited. Obviously, goggles for the correct wavelengths must be chosen, or they are of no value at all.

Other useful devices for laser safety include a shield over the apparatus, good enough to afford at least some seconds of protection from the direct beam. The human eye is unable to detect ir and uv hght. Therefore, lasers operating in these spectral regions offer a more serious hazard. Unsuspected reflections may occur. However, wraparound glasses or goggles may offer appreciable protection, even from tunable instruments, since the goggles may be made opaque to a wide range of wavelengths.

If any specific locations are particularly likely to suffer unwanted reflections, light traps, performing somewhat like Rayleigh horns,* may be installed. These are cheaper than Rayleigh horns, but, like Rayleigh horns, they cover only limited areas.

"Kill buttons," to turn off the beam remotely, may be carried by personnel in the room and used in case of emergency. Combined with a slightly transparent shield that affords some seconds of protection, these can be effective safety devices.

High power lasers require special precautions, beyond the scope of this discussion.

c / Microwave Radiation

The effects of microwaves are poorly understood but appear to be serious at fairly low dosages. Microwaves are absorbed by the body and produce heating effects. This is especially serious for the testicles and the lens of the eye. Testicles produce viable sperm only if they are below body temperature (presumably the reason for their external location). The lens of the eye is unable to lose heat, as it lacks a blood supply. Hence, the most obvious consequences of exposure to microwaves are cataracts of the eye and male sterility or possibly, at lower doses, birth defects in offspring. In addition, at still

---

*Horn shaped devices, optically non-reflecting, into which light enters, and is trapped.

lower doses, there are reports of damage to those organs that depend on electrical excitability, particularly the heart and nervous system. In particular, cardiac pacemakers fail under microwave radiation. If these reports prove correct, microwaves must be treated with even more respect than present standards suggest. Metal screening (mesh or plates) provides effective shielding against microwaves. If microwave ovens are in use, their closing mechanism must be in good repair if microwaves are not to leak.

The present United States standard for microwave radiation is 10 mW/cm² (milliwatts per square centimeter) power for any 0.1-hr period, 1 mW-hr/cm² energy, also averaged over any 0.1 hr period. This applies to whole body or partial body radiation (the present standard may be made more stringent, as there is evidence to suggest damage at these levels). There is a standard form for the required warning sign. A field strength meter is required for monitoring.

6.3 IONIZING RADIATION

We must be concerned with two varieties of hazards: external, such as that from x-ray sources, and internal, from the ingestion of isotopes.

a / External Radiation Standards

It is difficult to discuss a single standard for exposure to ionizing radiation. Different forms of radiation have different biological effects; the effect is also dependent on the target organ. The standard unit used is the **roentgen** (R), corresponding to the quantity of x or γ rays needed to produce 1 esu (electrostatic unit of charge) in 0.001293 g of air (1 cm³ at standard temperature and pressure). Doses of other forms of radiation, such as α or β rays, are adjusted so that they will have equivalent biological effect. Overall dosage is usually given in **rems** (roentgen equivalent man), to attempt to put the biological effects of all forms of radiation exposure into a single number. (Other units, such as the **rad**, equal to about 1.2R, are also used, but we will not discuss them here). The occupational standard is 5 rems/yr for whole body radiation, but 170 mrem is the population standard. It is estimated that a lifetime total exposure to

5 rem (not 5 rem/yr) is equivalent to a decrease in life expectancy of 1 to 3 weeks.

Work with x-ray instruments, or other external radiation sources (other than single exposures in undergraduate courses) should make provision for personal monitoring and for area monitoring if appropriate. Should excessive exposure be found, protective measures would be required, beginning with removal from further exposure.

b / Monitoring

Monitoring of personnel can be done with either of two types of dosimeters. First, pocket ion chamber dosimeters are about the size of a pen and have the advantage that they can be read at any time. They can be used to monitor dose and dose rate of γ rays, x rays, and hard (that is, high energy) β rays. Film badge monitors are more common. They consist of dental size films (about 3×4 cm) and are of two types, one of x-ray film, which shows general darkening on development proportional to total exposure to radiation, and another, which shows individual tracks of neutrons. The latter is important for those working with neutron sources, since the biological effects of neutrons can be more serious than those of other forms of radiation. For most purposes, the former type of film, which attempts to monitor the total dose of x, γ, and β rays, will be used. Since the response of the film to different energies of radiation varies, filters are used to try to compensate. Of course, if the exposure is to relatively monochromatic x rays, the problem is greatly simplified. Other problems with film include the effects of heat and aging. Nevertheless, film badges are normally the monitor of choice.

The film is developed after a set time, normally a month, and the total exposure estimated. Most laboratories will contract with an external firm to supply, develop, and read the badges.

i / Survey Monitors

These determine the dose rate in a given area, usually in milliroentgens per hour (mR/hr) or in disintegrations per minute (dpm). The most commonly used is a type of ionization chamber meter called the "Cutie pie" survey meter. It detects x and γ rays (8 keV to 2 MeV) with about 10 per cent error in the dose rate range 5000 to 10,000 mR/hr, and externally hazardous (high energy) β rays with about 50 per cent efficiency. Other common survey meters include the Juno survey meter, also an ionization chamber instrument, used as a relative-intensity meter for α rays. It has scales of 10^4, 10^5, and 10^6 dpm for α radiation. For β and γ radiation, it functions as a dose rate meter, with scales 50, 500, and 5,000 mR/hr. Manually positioned shields are used to discriminate among forms of radiation. (The γ rays penetrate somewhat farther than x rays, which in turn penetrate considerably more shielding than do β rays. Hence, shielding can be used to get a rough distribution of types of radiation.) The third common type is the Geiger-Muller (G-M) survey meter, based on the G-M tube as detector. This does not work well with β rays of less than 200 keV (unless a special thin window is supplied), but it can indicate radiation almost from background levels (perhaps 10 dpm) to well over 10,000 dpm. Efficiency with γ rays is low, however. In higher dose ranges, above about 50 mR/hr, this type of monitor saturates.

Other types of counters are less common or more specialized. Some are scintillation counters, some are sensitive to neutrons or α radiation. One that must be mentioned is a nonportable personnel monitor. This is a large instrument into which hands and feet may be inserted. If more than a set number of counts are detected, appropriate precautions should be taken. This is, however, a fairly expensive instrument, and is likely to be found only in larger installations. It should not be forgotten that if any danger does exist, it is necessary to have available the necessary equipment to deal with the hazard.

c / Isotope Laboratories

In general, it is to be hoped that any chemistry department that will handle radioactive isotopes, other than in sealed sources, will have obtained the advice of a competent health physicist or other similarly experienced person. For that reason, this section will be relatively brief. Unlike x-ray tubes, which can simply be turned off, isotopes emit continuously, and, if handled in systems which are not sealed, can be inhaled, ingested, spilled, or otherwise distributed in a dangerous and uncontrolled manner.

In the United States, local or federal licenses (depending on the location) are required for most laboratories using isotopes. Licenses generally are required for all unsealed sources; if the radioactivity does not exceed a few microcuries (μCi) they may be unnecessary— check with local or federal radiation control authorities. In the amounts sometimes required by chemists (up to a few microcuries) manufacturers will ship radioactive materials in sealed vials, labelled with the specific activity (in millicuries per millimole) or total activity (in microcuries). Those laboratories that use isotopes require a warning sign on their doors with the name and telephone number of the person to be called in case of emergency.

Obviously, what is written in the previous paragraph is for the benefit of the laboratories receiving the radioactive materials. Presumably the manufacturers exercise much more stringent supervision for the much higher levels of radioactivity in their laboratories and production facilities.

The laboratory itself must have storage areas, which may include a refrigerator and a hood. While "cold" (that is, nonradioactive) chemicals may also be stored in these areas, at least a section of the refrigerator or hood should be specifically set aside for the radioactive materials. Obviously, food should be kept far away from the radioactivity, not only from the storage area but from the entire laboratory.

In addition to the warning on the door, warning notices are also needed in the laboratory. They must include emergency procedures and are required in a conspicuous location in the laboratory.

Shielding requirements depend on the nature of the radioactivity; α radiation can be stopped by almost anything and will generally not penetrate its container. Low energy β radiation [which includes tritium (^{3}H or T) and carbon-14 (^{14}C)] can be stopped by plate glass or 1 or 2 cm of water. Hard β radiation or any γ radiation may require lead shielding. From a safety point of view, the requirement for shielding can be determined by use of a survey meter. The requirement for low background, if counters are in use in the same room, may actually be more stringent than the health requirement.

d / Protective Clothes
For work with low levels of radioactivity, gloves and a laboratory coat which may be disposed of if necessary, are required. The gloves in particular should be disposed of. During work, the gloves may

not be removed for any reason, nor may they be used to touch any apparatus or furniture that is to remain "cold," including drawer handles.

e / Training of New Personnel
An initial training session, including new research students, to insure that the nature and dangers of radioactivity are understood, is required. Routes of entry, including inhalation and absorption of fat-soluble isotopically labeled substances through the skin, must be discussed, as well as the possibility of absorption through small wounds. Rules of procedure and other information relevant to the safe handling of radioactive material in the particular laboratory, should be explained in detail.

f / Preparation of Experiments
A *complete* description of an experimental procedure must be worked out and written down before work is begun. This must include *all* materials to be used. It will not be possible, while wearing contaminated gloves, to open a drawer to pull out an extra beaker or stirring rod, and the gloves cannot be removed during the course of the work. The proposed procedure should be checked by one of the senior laboratory staff before it is carried out. The procedure must begin with the opening of the containers holding the labeled material and end with disposal procedures; it should also describe the layout of glassware and other materials to be used.

g / Confinement of Radioactivity
At the low levels being considered here, work may be carried out in an ordinary hood, provided the hood is in good condition, has adequate air flow, and there is no possibility of flow back into the laboratory because of air turbulence. Other considerations that make the use of a simple hood inadequate are use of particularly dangerous isotopes such as strontium-90 (^{90}Sr) or the production of excessive amounts of volatile products such as carbon-14 labeled carbon dioxide ($^{14}CO_2$) which might be exhaled by experimental animals. At higher levels of radiation, or if particular dangerous conditions exist, a closed system such as a glove box is needed.

The hood itself must be inspected for crevices or other small spaces that might accumulate dust and, therefore, radioactivity. All work must be done on absorbent material (such as diaper lining) backed with plastic, so that any spill will not spread. Before beginning work, *everything* for the experiment (all glassware included) must be on the absorbent-covered area.

If excessive volatile material will be generated (or other dangerous conditions such as those mentioned previously exist) a glove box should be used. Unfortunately, glove boxes leak. One way to collect most of the activity is to flow a gas (air, nitrogen, or whatever is appropriate for the experiment) through the box and absorb the gas flowing out in a train of gas-washing bottles. It may take as many as three bottles containing potassium hydroxide to absorb the $^{14}CO_2$ from a single experiment using mice. A better technique is to pump air *out* of the box and through the train so that air flow through any leak is into the box, not out. Of course adequate provision for make up air, or other gas, must be made, lest an unwanted vacuum be created.

h / Clean-up

A separate sink should be used for radioactive materials, including contaminated glassware. No more than 10 μCi (microcuries) (diluted by more than 4 liters of water) should be allowed to go down the drain, and then only of isotopes that present no particular biological hazard. Volatile materials, as mentioned previously, must be captured in liquid or solid form for disposal. A survey meter should be passed over gloves, lab coats, and other equipment, used during an experiment. If hot spots are found, they should be cut out and placed with the radioactive waste. Swipe tests, in which a piece of filter paper, absorbent tissue, or similarly absorbent cloth, is used to wipe a fixed area near the experiment and then counted, are mandatory as indicators of spill or other spread of radioactivity beyond the appropriate area. A test is also advisable from time to time for the floor near the radioactive work area.

i / Waste Disposal

A waste disposal company must be hired to remove radioactive wastes. They will provide appropriate containers. Some large con-

Table 6.1   UNITS OF ISOTOPE ACTIVITY

Unit	Abbreviation	Definition
Curie	Ci	3.7×10^{10} dpm; the number of disintegrations per minute in 1 g of radium
Microcurie	μCi	10^{-6} Ci
Picocurie	pCi	10^{-12} Ci
millicurie/millimole	mCi/mmol	10^{-3} Ci/10^{-3} mol; the *specific activity* of the isotope
microcurie/millimole	μCi/mmol	10^{-6} Ci/10^{-3} mol

taminated items may need to be placed in strong polyethylene bags until they can be removed and buried.

j / Badges
Film badges should be worn by all personnel working with radiation. (Sometimes pocket ion chamber dosimeters may be used.) See discussion in part (b) of Section 6.3 above.

k / Special Biological Hazards
Certain common isotopes are particularly hazardous, and require exceptional care in handling. Only a few are discussed here. (Table 6-1 gives the units in which isotope activity is measured.)

i /Carbon-14 (^{14}C)
This isotope has a physical half-life of 5730 years, and produces β radiation of 155 keV energy, sufficient to give it a range in the body of several cells or more. Biological half-life* depends on the compound into which the ^{14}C is incorporated. This period is roughly 10 days for the whole body average, 40 days for fat, but longer, possibly indefinitely so, for some lipids, which may, for example, be deposited in the brain. The environmental maximum permissible concentration (MPC) in air is 1×10^{-7} μCi/mL (soluble); MPC in

*The time for 50 per cent of the material to leave the body.

water, 8×10^{-4} μCi/mL (soluble). If incorporated into nucleotides, the decay of ^{14}C to nitrogen-14 (^{14}N) is in itself mutagenic. This isotope occurs naturally, and the normal body burden is of the order of 0.1 Ci.

ii / Tritium (T or ^{3}H)

With a physical half-life of 12.36 yr., ^{3}H decays more rapidly than ^{14}C, producing a very low-energy (18 keV) β radiation. For this reason, the radiation has a very short range, making this the isotope of choice in autoradiography. The biological half-life, as HTO (water with one H replaced by T), is 9.5 ± 4.1 days. Again, hydrogen in some organic compounds exchanges far more slowly, leading to a longer biological half-life. Because it appears in water, this isotope is easily dispersed, hence especially dangerous. MPC in air = 2×10^{-7} μCi/mL; in water = 3×10^{-3} μCi/mL.

iii / Strontium-90 (^{90}Sr)

This element follows calcium into bones and teeth, and, being concentrated, is especially hazardous. Its physical half-life is 28.1 years, but its biological half-life in bone is approximately 50 years. It decays via 0.54 MeV β radiation, but effective energy to bone is 5.5 MeV. MPC in air = 3×10^{-11} μCi/mL (soluble) and 2×10^{-10} μCi/mL (insoluble); in water = 3×10^{-7} μCi/mL (soluble) and 4×10^{-5} μCi/mL (insoluble).

These environmental MPCs are low enough to be of concern in experimental clean-up (10 μCi in 10 liters of water is 10^{-3} μCi/mL; thus, when working with particularly dangerous isotopes, one should exercise extreme caution in allowing *any* activity to go down the drain).

iv / Other Isotopes

Other isotopes that may be of special interest include potassium-40 (^{40}K), sulfur-35 (^{35}S), phosphorus-32 (^{32}P), and iodine-131 (^{131}I), as well as various metal isotopes. Of these four, ^{35}S is perhaps the easiest to handle. However, before beginning work, it is wise always to check the biological characteristics of the isotope to be used, as well as the chemical form(s) in which the isotope may appear because the latter may strongly influence the target organ of any radioactive substance that may be inhaled, ingested, or absorbed.

6.4 OTHER PHYSICAL HAZARDS

Noise, heat, cold, and other physical hazards are among the most common industrial problems (especially noise). However, they are not common in chemical laboratory work.

a / Noise
Noise is believed to be the most common industrial hazard in the United States at present. Above 85 decibels (dB) on the A weighted scale, it causes not only deafness in some workers but symptoms of stress, with corresponding effects on blood pressure and other organs. Above 90 dBA, the present 8-hr average standard, deafness occurs in many workers. In the laboratory, teletypes may be a problem, although several would be required to reach 85 dB. In general, exceeding the noise standard is not a problem in laboratories. If the annoyance effect is severe, either personal protective gear consisting of ear muff-like noise attenuators or engineering controls may be used. The engineering devices might be enclosure of the noise source or personnel, or setting the noise source at some distance from the personnel, or muffling the source itself. Engineering controls are far better, though more expensive in most cases (but sometimes a simple shim will do wonders with a rattle).

b / Heat
It is hard to see heat as a risk in a chemical laboratory in any direct sense (for example, causing danger of collapse). Conceivably some chemical engineering laboratories may encounter a problem. However, under more ordinary conditions, there is some danger to efficiency. Thermal stress progresses through six stages: discomfort, job inefficiency, continuous work with physiological risk, continuous work for specified duration, heat collapse, painful exposure. For any normal laboratory work, job inefficiency is already unacceptable, and ventilation or air conditioning must be provided. Studies have shown that sudden increases to as little as 23 to 25°C CET produce increases in accidents and absenteeism. CET = corrected effective temperature, defined as any environment "which produces an equivalent sensation of discomfort or warmth as an envi-

ronment which has an air temperature of the same value, 100 per cent relative humidity, and still air". Beyond 27°C CET, tasks that require fine motor manipulation, vigilance, and decision making are performed with deteriorating efficiency and accuracy. Less skilled operatives are more at risk than more skilled operatives. This may justify air-conditioning student laboratories in the summer. (Note that the limit is set for cases of little physical effort.)

c / Cold

There is obviously a loss of manual dexterity in the cold. Anyone working in a cold room must take this into account; one also requires suitable warm clothing. Other risks due to cold may involve the equivalent of burns—for example, handling dry ice without suitable gloves. Such silliness is in a class with horseplay, as the "accident" results from a conscious decision to go against the most rudimentary common sense.

6.5 SUMMARY

A variety of physical hazards may be encountered in chemical laboratories. Of these, ionizing radiation is probably the most serious and requires the most elaborate precautions. However, even apparently minor problems may, upon occasion, be matters for real concern, and some thought should be given to them.

REFERENCES

Ionizing Radiation
K. Z. Morgan and J. E. Turner (eds.): *Principles of Radiation Protection: A Textbook of Health Physics.* John Wiley and Sons, New York, 1967.
M. Eisenbud: *Environmental Radioactivity,* 2nd ed. Academic Press, New York, 1973.
C. W. Easley: *Basic Radiation Protection.* Gordon and Breach, New York, 1969.

7

Cleaning up the Laboratory

7.1 GENERAL PRINCIPLES OF CHEMICAL WASTE DISPOSAL

The problem of disposal of chemical wastes from laboratory operations involves difficulties of many kinds, not all of which have satisfactory solutions. Some laboratories retain questionable chemicals for many years, thus delaying the problem until some event such as remodeling of the building or change of laboratory function reveals it. A laboratory may hire the services of a qualified commercial organization to remove and dispose of its wastes; it may set up its own procedures; or it may utilize a combination of both facilities.

The selection of a commercial organization, however, does not automatically solve the problem, even if the costs are acceptable. For example, such organizations may exclude certain classes of wastes (such as alkali metals) from their services. Furthermore, regulations for packaging chemicals are under the jurisdiction of the Office of Hazardous Materials, Department of Transportation, Washington, D.C., and this agency should be consulted for its regulations concerning the preparation of wastes before they are turned over to the disposal company.

Most laboratories carry out their own disposal procedures. Three general methods are available: dilution, conversion to less hazardous substances, and "permanent" storage.

a / Dilution
In practice, dilution involves dumping chemicals into a receiving watercourse or sewer and releasing gases and vapors to the atmosphere. These procedures are widely used and abused. Discharge of toxic materials into a domestic sewage system may inactivate the organisms that perform the biological purification and thus magnify the resulting environmental pollution. Venting of waste gases to the atmosphere can sometimes be innocuous (at least on a global, if not a local scale). For example, methane, carbon monoxide, and hydrogen are natural components of the atmosphere in small concentrations, and mechanisms operate for their eventual cycling in the biosphere. Chlorinated vapors, however, such as carbon tetrachloride and dichloromethane (CH_2Cl_2) are very persistent, and may not be so benign. Nonetheless, it is difficult to dispose of them by any means other than venting.

b / Conversion to Less Hazardous Substances
The usual conversion procedures are neutralization of an acid or a base, reduction of an oxidizing agent, and, most common of all, oxidation by burning in the open air. In all such cases, the requirements for protection of the operator are most stringent, for the quantities being handled may be considerably greater than those involved in the usual laboratory operations. Open burning, particularly, may produce serious local air pollution.

c / "Permanent" Storage
This generally implies landfill, although the ocean has also been used as a repository in past years. A land site cannot be chosen casually, because consideration must be given to geological factors (especially the depth of the water table) and to effective protection against exposure of people, which implies a remote location and a fence.

7.2 SPECIFIC PROCEDURES

a / Compressed Gases
The first question to be answered in deciding on how to dispose of a defective cylinder of compressed gas is whether venting should be

slow or fast. The advantage of fast venting is that it affords the opportunity to choose a favorable time and, especially, a wind direction that will carry the gas away from populated areas. The method of choice is detonation by remote control of a sheet explosive taped to the outside of the cylinder. Added advantages are that there is no need to try to open defective valves and that the inadvertent reuse of the defective cylinder is prevented.

b / Solvents and Other Flammables

It is tempting to try to dispose of flammables by incinerating them with a homemade or commercially manufactured waste burner equipped with a suitable burner nozzle. The general experience with such procedures has been unsatisfactory, and we do not recommend it. The burning characteristics of mixed laboratory wastes vary over wide ranges, and the operation is often difficult to control. Furthermore, these wastes usually contain some chlorinated solvents, and the acids that result from their combustion corrode the burner and nozzle. If an isolated area is available (as for the detonation of gas cylinders), it is probably best to dig a pit there, dump the solvents in, together with any solid flammables, and ignite them with a railroad flare. The University of Minnesota reports that it disposes of about 25,000 liters of liquid solvent and 900 kg of solid chemicals per year in this manner. In the absence of such a facility, flammables should be turned over to a commercial disposal company. They should *not* be poured down the drain, not even in small quantities in student laboratories. (Trivial amounts of water-soluble solvents, such as alcohol or acetone, may be excepted.)

c / Explosives

An ancient bottle of isopropyl ether, a rusted can of yellowed sodium amide, a lump of partially oxidized potassium are examples of materials that are extremely dangerous to handle, let alone dispose of. Such hazards demand the utmost in personal protection for anyone who deals with them. It is by no means extreme to consider them to be unpredictable explosives to be handled only by professional bomb experts.

d / Toxic Materials

These may require a special set of procedures to insure against exposure of laboratory personnel before disposal or of anyone else afterwards. Small quantities of liquids or vapors can be conveniently adsorbed on activated carbon, which may then be bagged and disposed of in a landfill site. It is most convenient to use a granular, not powdered, carbon, in a carbon/adsorbate ratio of about 15:1 by mass. If toxic chemicals are to be burned, precautions must be taken to insure that no one is exposed to the smoke. If it is feasible to do so, a prior chemical conversion to a nontoxic or less toxic product should be carried out.

7.3 HOUSEKEEPING

There is nothing conceptually difficult about the method of keeping a laboratory clean and safe—it requires only an organized procedure and the will to carry it out. It is important to point out to laboratory workers, from novices to veterans, that good housekeeping promotes not only safety (which should be reason enough) but also more efficient performance. The shortcuts that sloppiness offers must eventually be paid for—the dirty equipment will have to be cleaned sometime, and the misplaced tool or chemical will eventually have to be found. These principles, however, must be implemented. If there is to be a limit on the quantity and storage time of specific chemicals, these limits should be specified in writing and posted or distributed to the laboratory workers. In student laboratories, clean-up time should be set aside as part of the laboratory program, not crowded into the few moments before (or after) the final bell. Finally, a schedule of inspection and checkup should be arranged to assure that good housekeeping is an integral part of the laboratory activity, not merely a set of rules on paper. (See also Sections 2.3 and 8.3).

A sample report form is shown in Figure 7.1.

_____ COLLEGE

Department of Chemistry

SAFETY INSPECTION REPORT

Room_____   Date_____ 19_____   Time of day _____

Inspector (s)_____

Function: Teaching Lab _____ Research Lab_____ Prep. and dispensing _____

          Storage_____ Other_____

ITEM	O.K.	None or not applicable	Uncorrected hazard (Describe)
SAFETY EQUIPMENT			
Fire extinguishers			
accessible?			
filled (see tag)?			
Sand/lime pails			
filled and clean?			
Gas masks			
accessible?			
recently tested?			
suitable type?			
Bicarbonate paste			
accessible?			
labeled?			
Hoods—working properly?			
Eye wash fountain—working?			
Safety shower-unobstructed?			

Figure 7.1   Safety Inspection Report.

Electrical equipment

    wires insulated?

---

    grounded?

---

    away from water or
    flame?

---

    wires where no one
    can trip?

Vacuum systems—set where
    protected against
    impacts?

Unattended equipment
    hoses wired to faucets?

---

    drain hoses secured
    at sink?

---

    water flow 1 liter/
    min?

---

    heaters shut off
    automatically if
    water stops?

Moving belts and wheels
shielded or set away from
aisle?

Figure 7.1 (Continued)

Compressed gases

      secured to desk or wall?

---

      capped if not in use?

---

      remote from radiators,
      ovens, burners?

---

      pressure reads zero if
      not in use?

Flammable liquids

      remote from heat or spark
      sources?

---

      in metal cans if in large
      quantity?

Corrosive or volatile chemicals—
stored in perfectly sealed
containers or in ventilated
cabinets under hoods?

Ethers

      date on label?

---

      free of solid deposits?

Cryogenic liquids—safely
      stored and vented?

Mercury

      NONE exposed anywhere?

---

      pans to contain mercury
      in case of breakage of
      manometers, diffusion
      pumps, etc.?

Figure 7.1 (Continued)

All chemicals closed and well
    labeled?

_____

    large bottles stored
    near floor?

_____

LABORATORY

Sinks unclogged; no debris?

_____

    faucets not leaking?

_____

Gas valves
    can be shut off tightly?

_____

    stop pins in place?

_____

Floor—not slippery or
    irregular?

_____

Cold room safety escape
    systems o.k.?

_____

General condition clean
    and orderly?

_____

RADIATION

_____

Recommended date for reinspection and other remarks

Figure 7.1   (Continued)

8

Administrative
Procedures

8.1 INTRODUCTION

The precautions and procedures outlined in the previous seven chapters are without value if they are not consistently applied. Consistent application requires the existence of a mechanism to ensure that laboratories conform to safe standards and that personnel are aware of, and apply, safe practice in their own work. To this end, certain administrative procedures are needed. In a typical department of chemistry in a university, a safety committee would be needed. An industrial laboratory may have a resident safety officer; even there, working chemists and technicians and other workers exposed to like hazards should be assigned to serve on a laboratory safety committee. Other administrative factors in safety include obtaining up-to-date information, insuring that appropriate safety instruction takes place, that accidents are reported, and that housekeeping procedures are in fact carried out. Maintenance of medical records may be the responsibility of such a committee.

8.2 SAFETY COMMITTEE

The success of a safety program depends on the group that is responsible for the administration of the program; this is the safety committee. In a university, the departmental safety committee should include faculty, technicians, and other workers. The latter groups are often most exposed to hazards, and most aware of poor house-keeping and maintenance and other hazardous conditions. Comments of workers without academic degrees must not be ignored because of false assumptions about professional expertise. Students may be included; the relative advantages or disadvantages of including students vary from place to place. The size of the safety committee may depend on the size of the department; each significant segment of the department should be represented.

The authority of the safety committee derives from the executive of the organization; a committee lacking the active support of the Department Chairman or Laboratory Director will almost certainly be ineffective. In general, the majority of the members of the committee are appointed by the executive, though this is not necessary. Any enforcement powers the committee may have derive entirely from the executive of the department or laboratory. For these reasons, a safety program that is weak or ineffective is also the responsibility of the head of the department or laboratory.

The responsibilities of the safety committee include the subjects discussed in the following paragraphs.

a / Obtaining Current Information
NIOSH and OSHA from time to time issue new criteria documents, regulations, and other circulars. (See the Bibliography, page 159). A safety committee should obtain current information, distributed by the local OSHA office, the Government Printing Office (Superintendent of Documents, GPO, Washington, D.C., 20402), and by NIOSH in Cincinnati. It is necessary to know which documents are being issued in order to obtain them. For this purpose, the regional OSHA office, such journals as the *International Journal of Occupational Safety and Health,* and, at much higher cost, various business service subscription newsletters may be consulted. A looseleaf book containing standards is maintained by OSHA, but there is no mailing list service.

Information on a particular substance may be obtained from NIOSH, 26 Federal Plaza, Cincinnati, Ohio. NIOSH criteria documents exist for a fair number of substances, and single copies may be obtained without cost from Publications, Division of Technical Services, NIOSH, 4676 Columbia Parkway, Cincinnati, Ohio 45226.

If the document stock number is known, it may also be ordered from the Government Printing Office (address in the preceding paragraph).

Form OSHA 20 contains a list of incompatible chemicals, toxicity information, and the manufacturer's suggested antidotes in case of poisoning with the particular substance to which it applies. Originally this form was supplied by the manufacturer and intended to be used for the transportation of the substance to allow safe clean-up in case of a spill. Form 20 is therefore used more generally than OSHA intended, but it is useful for the particular material covered.

It should be noted that OSHA will respond to requests for information from outside the United States, as well as from "governmental laboratories", as defined in Chapter One.

Other information is available from sources given in the bibliography. No single work can keep up with the flow of information, as new findings and new standards are continually issued. There is, obviously, useful information that is not contained in specific OSHA or NIOSH documents, as for example, in the safety columns in the *Journal of Chemical Education.*

b / Accident Reports

All laboratory accidents should be reported to the safety committee. An accident may be as small as a cut finger from broken glass in a freshman laboratory; it may be a spill of toxic or carcinogenic material. Accidents are not limited to incidents resulting in overt serious injury. The safety committee should ask that a standard accident report form be filled out. Based on the information, it may conduct a further investigation or simply add the accident to statistics that it should be keeping on types of accidents in the department or laboratory. The statistics show whether the frequency of certain types of accidents is excessive, and whether they might be eliminated or made less frequent by suitable improvements in procedures or training. A possible form to use for accident reports is shown in Figure 8.1. *Note:* in addition, a report of this sort could also be used for legal purposes.

c / Maintaining Health Records

OSHA rules require periodic medical examinations, tailored to the specific hazard (for example, for exposure to benzene, one test is

REPORT

Date and time of report _____

Your name _____

Date and time of accident _____

Class and room number _____

Location of accident in room _____

Where were you at the time? _____

What were you doing? _____

Describe what you observed _____

_____

What did you do after the accident occurred? _____

_____

Any other remarks? _____

_____

_____

Signed _____

Figure 8.1   Accident report form.

for phenol in the urine). Medical records must be kept for many years after exposure has ended (hence, usually after the employee has retired). At the time of writing, OSHA enforcement is just beginning, although these rules apply to universities (except public universities) as well as to industry. In the absence of monitoring, the rules appear to apply to anyone in the vicinity, hence possibly to the entire department for each of many substances. It is hoped that a combination of care in monitoring and confining carcinogens and exercise of reasonable judgment will limit the number of persons considered to be exposed to each substance. The safety committee may not be the appropriate administrative body to maintain health records; it should nevertheless be sure that the records are maintained.

d / Other Recordkeeping:

i / Occupational

Forms for recording injury are prescribed by OSHA, and have been simplified as of January 1, 1978. OSHA rules require the use of one form, OSHA 200, shown in Figure 8-2. OSHA 200, the injury and illness log, must be maintained for 5 years. If there is a change of outcome or extent of injury or illness, such as an employee returning to work and later dying of the illness, the original entry is to be lined out and the corrected information entered. The forms are contained in a booklet issued by OSHA, "Recordkeeping Requirements Under the Occupational Safety and Health Act of 1970"; the booklet can be obtained from regional OSHA offices.

The injury and illness log section of Form 200, giving totals but no individual data, must be posted no later than February 1 of the year following that for which the log has been kept, and must remain posted until at least March 1. (OSHA allows the substitution of forms containing all equivalent information for the forms which it distributes.)

ii / Non-Occupational

While there is no comparable legal requirement for keeping records of student injury and illness, it seems reasonable to expect that similar records would be of value to any departmental safety program. They would make possible a ready determination of patterns of injury or illness, and aid in locating non-obvious hazards.

In addition to these records, medical records are required in case of exposure to chronically toxic substances, such as benzene. It may be necessary to maintain medical records for an extended period (for example, 20 years in the case of benzene).

e / Housekeeping Procedures-Control

Housekeeping procedures have been described in Chapters 2 and 7. However, it is necessary to insure that the appropriate procedures

Bureau of Labor Statistics
Log and Summary of Occupational
Injuries and Illnesses

NOTE: This form is required by Public Law 91-596 and must be kept in the establishment for 5 years. Failure to maintain and post can result in the issuance of citations and assessment of penalties. *(See posting requirements on the other side of form.)*

RECORDABLE CASES: You are required to record information about every occupational **death**, every nonfatal occupational **illness**, and those nonfatal occupational injuries which involve one or more of the following: loss of consciousness, restriction of work or motion, transfer to another job, or medical treatment (other than first aid). *(See definitions on the other side of form.)*

Case or File Number	Date of Injury or Onset of Illness	Employee's Name	Occupation	Department	Description of Injury or Illness
Enter a nonduplicating number which will facilitate comparisons with supplementary records.	Enter Mo./day.	Enter first name or initial, middle initial, last name.	Enter regular job title, not activity employee was performing when injured or at onset of illness. In the absence of a formal title, enter a brief description of the employee's duties.	Enter department in which the employee is regularly employed or a description of normal workplace to which employee is assigned, even though temporarily working in another department at the time of injury or illness.	Enter a brief description of the injury or illness and indicate the part or parts of body affected.

Typical entries for this column might be: Amputation of 1st joint right forefinger; Strain of lower back; Contact dermatitis on both hands, Electrocution—body. |
| (A) | (B) | (C) | (D) | (E) | (F) |

PREVIOUS PAGE TOTALS →

TOTALS (Instructions on other side of form.) →

OSHA No. 200
GPO 960-502

Figure 8.2 OSHA Form 200, Log and Summary of Occupational Injuries and Illnesses (left part of the form).

U.S. Department of Labor

For Calendar Year 19 ____   Page ____ of ____

Form Approved
O.M.B. No. 44R 1453

Company Name

Establishment Name

Establishment Address

Extent of and Outcome of INJURY | **Type, Extent of, and Outcome of ILLNESS**

Fatalities	Nonfatal Injuries				Type of Illness							Fatalities	Nonfatal Illnesses				

Extent of and Outcome of INJURY

Fatalities — Injury Related

Enter DATE of death. Mo./day/yr. (1)

Nonfatal Injuries — Injuries With Lost Workdays

(2) Enter a CHECK if injury involves days away from work, or days of restricted work activity, or both.

(3) Enter a CHECK if injury involves days away from work.

(4) Enter number of DAYS away from work.

(5) Enter number of DAYS of restricted work activity.

Injuries Without Lost Workdays

(6) Enter a CHECK if no entry was made in columns 1 or 2 but the injury is recordable as defined above.

Type, Extent of, and Outcome of ILLNESS

Type of Illness

CHECK Only One Column for Each Illness *(See other side of form for terminations or permanent transfers.)*

(7)
(a) Occupational skin diseases or disorders
(b) Dust diseases of the lungs
(c) Respiratory conditions due to toxic agents
(d) Poisoning (systemic effects of toxic materials)
(e) Disorders due to physical agents
(f) Disorders associated with repeated trauma
(g) All other occupational illnesses

Fatalities — Illness Related

(8) Enter DATE of death. Mo./day/yr.

Nonfatal Illnesses — Illnesses With Lost Workdays

(9) Enter a CHECK if illness involves days away from work, or days of restricted work activity, or both.

(10) Enter a CHECK if illness involves days away from work.

(11) Enter number of DAYS away from work.

(12) Enter number of DAYS of restricted work activity.

Illnesses Without Lost Workdays

(13) Enter a CHECK if no entry was made in columns 8 or 9.

Certification of Annual Summary Totals By _____ Title _____ Date _____

OSHA No. 200

FOLD

POST ONLY THIS PORTION OF THE LAST PAGE NO LATER THAN FEBRUARY 1.

Right-hand part of OSHA Form 200.

149

have in fact been carried out. This can be accomplished by inspection of the laboratories, which in turn can be done by the individual responsible for the laboratory or by the safety committee. The choice may be a matter for judgment by the safety committee; at least the possibility of a check by the committee should be maintained. The results must be recorded on an appropriate form such as that shown on pages 139–142.

Another aspect of housekeeping for a department or laboratory with a number of researchers is keeping track of hazardous chemicals and carcinogens. A form for listing the amounts, location, and type of storage of carcinogens and other seriously toxic substances should be filled out for each such substance, by each person using or storing it. Because of the fact that college laboratories typically use carcinogens, such as benzene, the safety committee should be in a position at least to suggest substitutions in student laboratories; in some cases the department may even want to allow the committee to be able to order substitutions. Furthermore, the committee should keep track of the use of carcinogens in research laboratories.

The decision as to which chemicals are in need of rigorous control is itself a matter of judgment to some extent. For example, vinyl chloride is a known carcinogen. No standard has yet been issued for vinyl bromide. Nevertheless, given the general tendency of organobromine compounds to be more toxic than their organochlorine analogs, it would be foolish to leave vinyl bromide uncontrolled. (Evidence that vinyl bromide is a carcinogen has in fact recently appeared.)

Standards are given throughout the book for exposure to physical and chemical hazards. At several points we have found it necessary to caution against taking these standards as absolute indicators of safety. Some older standards were developed as protection against acute toxicity only. In the absence of appropriate testing, standards are often not set, or are set on the basis of very limited data. Generally, additional testing finds the standards not to be sufficiently stringent. An important reason for standards not to be sufficiently stringent is the requirement of "economic feasibility"; if a standard is alleged to involve such high costs that it may threaten the economic viability of the industry, there is a great tendency to relax the standard, regardless of other considerations. Also, when a strict standard is promulgated by a government agency, its legal adoption and enforcement may be delayed, sometimes for years, by litigation that centers around the question of economic feasibility. (For example, at present the benzene standard is being challenged in this manner.) For these reasons, when new evidence on the health effects of toxic chemicals becomes available, the usual finding is that the standards have in fact been too lax rather than too strict.

f / Summary

Safety committees should represent the entire laboratory or department, including technicians, dishwashers, stockroom employees, and so on, and possibly students in a university laboratory. The committee has the task of seeing to it that rules are enforced and that proper housekeeping procedures are followed. Accidents must be reported to the safety committee; it has the responsibility for investigating them and keeping appropriate accident records. The committee must also be sure that proper medical records exist for employees exposed to hazardous substances. Finally, it must keep track of the hazardous or controlled substances.

8.3 TRAINING OF EMPLOYEES, NEW INSTRUCTORS, AND STUDENTS

a / OSHA Requirements

The OSHA requirement for training employees exposed to a given substance depends on the standard for that substance. Usually, those workers requiring training are those who may have significant exposure, which is generally defined as half the TLV or TWA. Presumably, OSHA will eventually issue regulations appropriate to laboratories, as opposed to production situations. For those laboratories that are in a position to create their own regulations ("government" labs in the United States, and laboratories in some other countries), it may be sensible to instruct all incoming staff, professional and nonprofessional, in general principles of handling toxic substances. If all substances are properly labeled and written instructions are available for toxic materials, personnel may be told to consult the written instructions for each material; these written instructions may include such data as are found in Form OSHA 20, plus anything prepared specifically for the particular laboratory.

The content of the instruction is essentially that which has been outlined in previous chapters with respect to handling materials, housekeeping, safe disposal, and so on. Anyone using particular hazardous materials should be listed with the safety committee. If new information becomes available concerning a substance, the safety committee may inform those using the material and even carry out retraining, if necessary.

b / Training of Instructors in Universities

Instructors are responsible not only for their own safety but for that of their students as well. Therefore, they must not only be well enough prepared to choose safe procedures in the laboratory but also to instruct students in these procedures. It is to be hoped that eventually all chemistry graduate students will be as well informed about safety as about any other subject they may be called upon to teach. At present, however, graduate students are frequently assigned to teaching duties without having gained adequate information about the hazardous properties of many substances, and without systematic instruction in safe procedure. Fortunately it does not take a great deal of time to learn enough to conduct a student laboratory safely (provided a senior staff member is available should difficult problems arise).

The department must insist that new instructors show that they have the necessary acquaintance with health and safety problems. Instructors must also learn basic emergency procedures, such as the use of respirators. If the course uses particularly hazardous substances, such as benzene or carcinogenic aromatic amines, it is necessary to emphasize the necessity for such precautions as protective clothing, proper housekeeping in case of spills, and so on. Presumably, the use of these substances will eventually be reserved to more advanced courses, in which their safe handling will be an integral part of the curriculum. If this is done, special training for teaching assistants will become less necessary, and chemistry graduates will have already received much of this training as undergraduates.

c / Teaching Undergraduates

Undoubtedly each school will choose its own method. It is our feeling that safety instruction is best integrated with regular laboratory courses. However, one of the principal difficulties is persuading students of the necessity for safety precautions. Students persist in removing glasses or goggles in the laboratory and otherwise show that they do not understand the seriousness of the risks involved. It may well prove worthwhile to take a couple of hours, either during the first laboratory session (especially if students would otherwise be sent home early) or from the lectures associated with the laboratory, to discuss the hazards of the particular laboratory course and the general rules of safe laboratory work. If new hazards (say, new substances) are introduced later in the semester, their associated safety problems may be discussed at that time.

A new teaching assistant, faculty member, or laboratory worker may already be familiar with basic principles of safe practice and know how to teach, but nonetheless must be taught how to teach these principles of safe practice to others. The first requisite is that the new person should be made to feel confident that the department or laboratory has a real commitment to safety, which means that regular classroom or laboratory time should be devoted to such instruction as an integral part of the teaching or research function.

Second, the newcomer should be assured that it is necessary to teach safety to students or other novices *at an elementary level.* Every experienced teacher who has seen students burn themselves with hot glass they have just withdrawn from a flame understands this only too well, but the new young teacher may feel that such danger is so obvious that it would be patronizing to point it out. It is not patronizing, it is necessary. Other such "obvious" precautions as the need for carrying a bottle of concentrated sulfuric acid firmly by its cleaned body, rather than gingerly by its untested cap, also need emphasis. It is hardly possible to overdo such instruction. In any event, some student is likely to discover a new hazard* that never occurred to the instructor.

Third, the new staff member must be told that instructions alone are not enough. In a teaching laboratory, all equipment setups must be checked. This means that the instructor must examine each setup just prior to the last, crucial step, which is usually the input of heat

---

*For example, a student complained to the instructor, "Every time I try the brown ring test for nitrates, my fingers burn!" The instructor knew that the procedure called for pouring concentrated sulfuric acid "carefully down the side of the test tube," but never dreamed that the student would think that meant the *outside*! Every experienced teacher knows such sad stories.

or the addition of the final reagent. In a distillation apparatus, for example, the instructor must check for mechanical stability, integrity of connections, venting to atmosphere, flow of cooling water, presence of boiling aids, and avoidance of an overfilled flask before giving the student permission to apply heat.

Finally, the instructor must be made aware of the sensitive balance between students' rights and responsibilities. Long hair must be kept away from a flame by being confined or tied back. It is unlikely that anyone will object to such a requirement, but the gender of the student is irrelevant and should not be commented on. Similar considerations apply to protective clothing. An instructor who finds that a student is working dangerously or is engaging in horseplay must see to it that such actions are immediately brought to a halt, and may properly regard such behavior as a lapse in academic performance. Questions of disciplinary action after the hazard is over, however, are to be handled in the same way as the institution provides for any other disciplinary matters.

The instructor will find that an initial statement of a very firm commitment to safe practice in the laboratory will be greeted neither with protest nor approval but with silence. However, under that silent facade most students will be secretly relieved that their safety is a matter of serious concern.

8.4 CONTINUING EDUCATION IN SAFE PRACTICE

The preceding section has dealt largely with administrative policy regarding the indoctrination and training of novices, particularly students. Education in safe practice, however, is by no means a closed-end subject. New laboratory procedures, new devices, and new chemicals all involve new potential hazards, and safety training must therefore be regarded as a continuing process, even for experienced workers. There is a strong tendency for interest in such programs to lag, especially when they are successful.

After all, the effective administration of safety policy is one which, by definition, results in elimination of spectacular events and, therefore, comes to be regarded as dull and, worse yet, intrusive. The executive body of an academic department or an industrial laboratory must therefore exercise its authority to maintain a continuing program of safety education and practice, and the safety committee must recommend or provide the needed information and materials. Some suggestions are given in the following paragraphs.

a / Safety Seminars

A panel discussion, or lecture-demonstration, open to all, should be scheduled at least once each year, perhaps as an integral part of the regular departmental seminar. Outside speakers may be invited or representatives of firms that manufacture safety equipment may demonstrate their products, or a representative of the Safety Committee may report on conditions in the institution's own laboratories.

b / Safety Library

A repository of safety handbooks and current literature should be available for reference to all. It should be accessible during every working day, but not necessarily at night or during weekends. This literature should be regarded as a resource of information to help in planning accident prevention, not as a place to which to run during an emergency. The department office, or the office of the Safety Chairman or Director of Laboratories are reasonable locations for the safety library.

c / Dissemination of New Information

As discussed previously, new literature on safety appears continually. The Safety Committee should collect all such material that it can obtain and circulate notices periodically to all staff members of the new information available in the safety library.

d / Safety Quizzes

Any examination is a challenge and therefore commands more attention than simple exposition, as editors of the puzzle pages of newspapers are well aware. The best questions are based on real problems, not contrived ones. For example:

You work in a laboratory near a mining town in Pennsylvania, and discover to your horror that a gas cylinder containing hydrogen at 1000 lb/in² has been pressurized with air to a total pressure of 2000 lb/in². The cylinder is strapped to the work bench and the main valve is shut. What would you do? (See answer at the end of this section.)

Questions, preferably based on real experiences, should be solicited from staff members, especially those who have worked in other laboratories. Other questions may be gleaned from literature reports or other sources and circulated with requests for answers. The various answers, including the "correct" one (if there is any) should then be distributed. If laboratory reports are required, one section might be added to deal with safety problems of the particular experiment.*

It is surely no secret to the reader that the primary purpose of these exercises is to maintain a high level of awareness about safety among all staff members and that the actual dissemination of information is secondary. Do not underestimate the challenge of maintaining a *permanent awareness* of safe practice. The effort will not succeed unless the administrators and safety officials bring a strong personal commitment to it.

8.5 REPORTING ACCIDENTS—LEGAL QUESTIONS

An accident report does not directly aid the victim, although the details it provides sometimes do help to guide the course of medical treatment. It is hoped that the accumulated information from accident records over a period of years will suggest preventive measures, and in this way any one accident report adds to the safety of other laboratory workers. If such purposes were the only ones, accident reports would be solely technical and clinical. The identity of the person in charge of the laboratory, for example, would be irrelevant.

Of course, this picture is unrealistic, for it ignores the requirement for preserving evidence that may be needed for a legal defense. The

*One way to do this is described by J. W. Hill and R. J. Pasteris in "A Modernized Mole Table," *J. Chem. Educ.* **52**:A291 (1975).

chemist who has never been involved in a legal case will be as much a stranger in court as the lawyer would be in the laboratory. It is therefore foolhardy for the chemist to make inappropriate assumptions about legal matters such as occur, for example, when a teacher who has had excellent relationships with students takes it for granted that an injured student will not sue. On the contrary, malpractice suits have never been more common, and the teacher or scientist is by no means immune.

The general advice from lawyers is this: gather and preserve all the evidence you can, but avoid any actions or statement that are the prerogatives of a judge or jury. Needless to say, make no attempt to "shape" the evidence in any way—this would be morally wrong and legally foolish. These caveats are interpreted so rigorously by lawyers that one of them* urges that an injury-causing event be called an "incident," not an "accident," because the word accident implies fault and evokes sympathy. He suggests that four types of evidence be preserved: testimonial, photographic, objective, and documentary.

Testimonial evidence from observers such as student witnesses should be obtained as soon as it is reasonable to do so, preferably on the same day as the accident (or incident, if you prefer). A suggested form in shown in Fig. 8.1.

Students should be told not to disturb the scene of the accident. After they have left the laboratory, photographic and objective evidence may be gathered. If a camera is not available, a sketch of the scene will be helpful. Don't forget the blackboard, which illustrates what was being taught at the time. Objective evidence includes remainders of equipment and reagents, residues scraped from the benchtop, and so on. They should be bagged or bottled, labeled with date, time and description, and preserved under lock and key under the protection of a *neutral* party (not a school administrator). Finally, gather and preserve documentary evidence such as your roll book, instructional material (especially with regard to safety), labels and instructions that accompanied the chemicals or equipment the student was working with, and other information regarding the experiment that was involved in the accident.

---

*James R. Gass, in *Safety in the Chemical Laboratory,* Vol. 3. Journal of Chemical Education Reprints, Easton, Pa., 1974, p. 8.

Answer to Safety Quiz on Pages 155-156

An accident such as the one described did occur some years ago at the Bruceton, Pa., laboratories of the U.S. Bureau of Mines. A number of cylinders were involved. The first was bled out slowly with the aid of the reducing valve on the automatic pressure regulator. An attempt to do the same with a second cylinder resulted in an explosion that caused two deaths. It was later speculated that the warming of hydrogen gas on expansion may have set off the explosive mixture. The remaining cylinders were then taken, with careful grounding, into a mine, and dynamited, one by one.

Bibliography

I. OSHA literature, available from Superintendent of Documents, U.S. Government Printing Office, Washington D.C. 20402.

General Industry Standards (OSHA 2077; also known as 29 CFR 1910). This is absolutely necessary; a 930 page book containing the basic legal safety standards.

Questions and Answers to Part 1910—OSHA 2095. Free.

Training Requirements of OSHA Standards. Free.

Occupational Safety and Health Subscription Service (issued in a three-ring binder; order forms may be obtained from the nearest OSHA office):

	Annual rates:
General Industry Standards and Interpretations	$21.00
Maritime Standards and Interpretations	6.00
Construction Standards and Interpretations	8.00
Other Regulations and Procedures	5.50
Field Operations Manual	8.00

Job Safety and Health (magazine). Includes news from NIOSH and from the Occupational Safety and Health Review Commission and a listing of the latest Federal Register insertions. $13.60 a year, monthly.

Job Health Hazards
Carcinogens (OSHA 2220). On the 14 original members of the OSHA carcinogen list. 50¢.
The following are free from OSHA:
Vinyl Chloride (OSHA 2225)
Carbon Monoxide (OSHA 2224)
Lead (OSHA 2230)

Mercury (OSHA 2234)
Beryllium (OSHA 2239)
Toluene Diisocyanate (TDI) (OSHA 2248)

Record-keeping requirements under the Occupational Safety and Health Act of 1970. Includes OSHA form 200 and instructions for its use, including legal requirements. Free.

OSHA Publication and Training Materials (OSHA 2019). Includes a complete list of OSHA publications, training materials, guidelines for compliance, and so on. Free.

II. NIOSH Publications. A list is available from: Publications, DTS, NIOSH, 4676 Columbia Parkway, Cincinnati, Ohio 45226. These include health and safety guides for various industries, results of specific studies, criteria documents, and sampling and analytical techniques for 16 classes of compounds, each with five to 16 compounds.

One document of particular importance is *Suspected Carcinogens—A Subfile of the NIOSH Toxic Substances List,* H. E. Christiansen and T. T. Luginbyl, editors, and B. S. Carroll, project coorinator. Available from Government Printing Office.

A second publication which belongs in *every* safety library is NIOSH Publication No. 77–181, *Occupational Diseases, A Guide to their Recognition* (Revised edition, June 1977). This work is primarily intended to alert health personnel to occupational illnesses. It contains almost 400 pages on chemical and physical hazards, as well as the most thorough list of reference aids available.

III. Publications of the American Industrial Hygiene Association (AIHA), 66 South Miller Road, Akron, Ohio 44313. (This is not a government organization.)

Analytical Guides for 61 Substances. $12.75 to AIHA members, $15.75 to nonmembers.

Basic Industrial Hygiene Manual. On monitoring, ventilation, etc. $8.75 to AIHA members, $10.75 to nonmembers.

IV. National Fire Protection Association, 470 Atlantic Avenue, Boston, Mass. 02210

Standards for Combustible Materials

V. Other useful information, either on toxic materials and procedures or on OSHA.

Encyclopedia of Occupational Safety and Health, International Labor Office, Geneva. 2 volumes. McGraw-Hill, New York, 1971.

N. I. Sax, *Dangerous Properties of Industrial Materials* (Van Nostrand Reinhold, New York, 1975)

Merck Index, 9th ed. (Merck & Co., Inc. Rahway, N.J., 1976)

J. Stellman and S. Daum, *Work is Dangerous to your Health.* Pantheon, New York, 1973.

Safety in the Chemistry Laboratory. Collections of articles from the safety column of *Journal of Chemical Education.* (Division of Chemical Education, American Chemical Society).

Safety manuals of certain companies are available, for example, those of Smith Kline Corp. or the Honeywell Research Center.

VI. Periodicals

Job Safety and Health (From OSHA; described above).
Journal of the American Industrial Hygiene Association (from AIHA).
International Journal of Occupational Safety and Health Occupational Hazards (PO Box 5746–U, Cleveland, Ohio 44115).

Index

Additional information on individual substances may be found in Table 4.1, pages 43–44 (Properties of Flammable Substances), Table 4.2, page 51 (Peroxidizable Compounds), and Tables 5.1, 5.2, and 5.3, pages 68–80, which list allowable exposures to various substances.